FAR BEYOND ZERO

THE TRUE PATH TO SAFETY CULTURE

Andreza Araújo

General coordination

Andreza Araújo

Cover

Jonatas Costa

Graphic Design and Layout

Reginaldo Saw

International Cataloging in Publication Data (CIP)

(Brazilian Book Chamber, SP, Brazil)

Marques, Andreza Moleiro Araújo

Far beyond zero : the true path to Safety Culture / Andreza Moleiro Araújo Marques. -- Barueri, SP : Ed. da Autora, 2025.

186p.

ISBN 978-65-01-47844-9

1. Work accidents - Prevention 2. Organizational culture 3. Engineering 4. Leadership 5. Safety - Measures

I. Title.

25-273339 CDD-363.1

Indexes for systematic catalog:

1. Safety : Measures : Social problems 363.1

Eliane de Freitas Leite - Librarian - CRB 8/8415

Estamos juntos nesta Jornada!

Uma Cultura de Segurança melhor para todos.

PREFACE

We are going through a decisive period in the history of organizations. The speed of technological, social and economic change is growing exponentially, bringing unprecedented levels of complexity to production systems. In the midst of this fast-paced scenario, the pressure for results has never been stronger, placing the workplace safety at a crucial and decisive moment.

For decades, our vision of safety was based on what seemed like a solid foundation, which was the absence of work accidents. We effusively celebrated milestones such as "zero work accidents" and counted the days without any injuries. Entire cultures were built on the simplistic idea that safety only meant the non-occurrence of undesirable events. However, this perspective proved to be insufficient and, in many cases, even dangerous.

The concept "Far Beyond Zero" urgently calls for a deep reflection on the way organizations perceive and practice safety. This is not just a slogan or a passing initiative, but a consistent call for cultural change. We are talking about transcending the merely reactive mentality, one that addresses safety only after damage has occurred, to evolve towards a proactive vision, developing capabilities even before problems arise.

Recent research, such as McKinsey's, reveals a worrying reality when it indicates that around 92% of organizations do not significantly evolve their safety cultures. They remain stuck in old behavioral patterns, creating a false sense of safety that becomes devastating in the face of real and unexpected events. We need to ask ourselves how we can evolve this stagnant culture.

This cultural paralysis does not stem from a lack of resources or investment in equipment, training or technical procedures. Many companies with chronic safety challenges have sophisticated technical systems, comprehensive policies and rigorous training programs. The real difficulty lies in something much deeper, invisible and subjective, which are the ingrained beliefs and values that govern employees' daily routines.

Far Beyond Zero is not implemented with a simple memo or a set of new rules posted on the walls. It must function as a living organism, fed daily by leadership decisions, trust and continuous learning. It is not something that is imposed, but something that is built through consistent practices and daily interaction.

This book is the result of more than twenty years of practice with various organizations around the world, the fruit of intense dialogues with CEOs, managers and workers who face the daily challenge of balancing pressure for results and human responsibility. This experience has made it clear that an organization can only achieve higher levels of safety if it deeply engages its leadership and all its teams.

The marathon metaphor often used in this context accurately reflects this long-term journey. Just as a marathon runner doesn't win with just the initial effort or quick, one-off runs, an organization doesn't achieve safety excellence with isolated campaigns alone. The marathon requires meticulous preparation, a consistent strategy and formidable mental stamina. In the same way, sustainable safety requires continuous efforts, flexibility and a deep vision, understanding that the path is as relevant as the goal.

Along the way, we will uncover the various layers that make up the concept of Beyond Zero, understanding unconscious assumptions that underpin organizational behavior and exploring the four levels of cultural maturity, which are reactive, dependent, independent and interdependent.

Reaching this level is not an unattainable ideal; on the contrary, it is a concrete and urgent possibility, within the reach of any organization if there is courage, persistence, humility and flexibility. The fruits of this transformation are unquestionable, including more resilient organizations, more engaged employees and safer, more prosperous communities.

The marathon is starting, and the route will certainly be challenging. However, the finish will be transformative and rewarding.

Welcome to this extraordinary journey beyond zero.

THANKS

Writing a book like "Far Beyond Zero" is an exercise in humility. I recognize that the words gathered here are the fruit of countless conversations, shared experiences and lessons learned alongside extraordinary people who dedicate their lives to making the world of work safer and more humane.

My gratitude extends to the visionary leaders I've had the privilege of meeting along this journey. They are CEOs, directors, managers and supervisors who have understood that safety is not a cost, but an investment. They have understood that taking care of people is not just a moral responsibility, but a sustainable competitive advantage. Their courageous decisions have lit the way for more resilient and prosperous organizations.

I would like to thank the safety, hygiene and occupational health professionals, engineers, technicians and specialists who work tirelessly behind the scenes on prevention. You are the silent architects of safer environments and the vigilant guardians who anticipate risks and build defenses. Your technical dedication combined with your genuine passion for people's well-being is a constant source of inspiration.

To the people on the front line, operators, technicians and employees from all functions who generously shared their experiences, fears, frustrations and hopes. You are the true protagonists of this story, because your authentic voices, your practical suggestions and your courage to talk about what really happens daily have given life and credibility to every page of this book.

I would also like to thank the researchers, academics and thinkers who have dedicated their careers to the study of organizational safety. Their rigorous research, innovative theories and scientific publications have provided the solid foundation on which these ideas could be built.

To my family, who understood the long hours of writing, the trips for case studies and the time dedicated to this mission. Your unconditional support and loving patience made it possible to turn years of experience into organized words.

To the organizations that opened their doors to allow case studies, diagnoses and the implementation of transformation programs. Their transparency and willingness to learn, even while exposing vulnerabilities, demonstrate the maturity needed to evolve.

I especially recognize those who, unfortunately, have learned through their losses. Each work accident, each interrupted life and each bereaved family reminds us of the urgency and importance of this mission. Their stories illuminate the need to keep moving forward and never settle for the status quo.

This book is the result of a collective construction, of a network of people who believe that it is possible to go "Far Beyond Zero". It is possible to build organizations where safety is a lived value, where care is mutual and where each person can return home with their integrity preserved.

To all of you, my eternal gratitude.

Andreza Araújo

INTRODUCTION

THE INEVITABLE JOURNEY TOWARDS SAFETY CULTURE

In a world where business is changing with speed, uncertainty and increasingly intense connections, safety is no longer just an operational obligation or a set of standards to be followed. It has become a strategic foundation, an essential basis for the continuity, credibility and true success of any organization.

Throughout my career, working with experienced leaders from a wide range of industries and countries, it has been possible to see the multiple layers that make up a consistent safety culture. In every conversation, from CEOs to people on the front line, one thing became clear: safety is no longer an isolated sector or a mere list of mandatory tasks. It is a value that pulses at the heart of organizations, shaping decisions, guiding attitudes and manifesting itself in both broad strategies and small everyday gestures. It is, above all, a profound way of thinking, acting and existing.

But why write a whole book about it, especially now? The answer lies in a truth that, although uncomfortable, insists on knocking on our door. Even in the face of so many advances in technology, engineering and management systems, many organizations continue to face real and persistent difficulty in making safety culture evolve in a consistent and lasting way.

Many organizations are still stuck in cycles of reaction, running for cover only after an incident has occurred. Instead of cultivating, with intention and care, the conditions that would prevent future incidents, they end up hostage to their own urgency. Others get lost in processes and controls that are so complex that they end up hindering, rather than facilitating, the safe execution of their work. They create cumbersome and bureaucratic systems, distancing people when they should be bringing them together and engaging them.

Recent McKinsey studies show that around 92% of organizational cultures simply don't move forward when it comes to safety. They don't transform, they don't mature, and they don't move. And that is not just frustrating, it is dangerous.

When the rush for results takes the place of purpose, when production is worth more than life, when there is no clear model of leadership and governance, what remains are improvisations, shortcuts. It is the "normalization of deviation", where the wrong thing is accepted because "it is always been this way".

The result is a painful cycle of frustration and disengagement. When this cycle is not broken, the price paid is extremely high. There are work accidents that maim, emotional traumas that sicken, pain that arrive without asking permission and take away what matters most, the precious integrity of a life.

That is why you can't just treat safety as a one-off goal. It needs to be cultivated as a lifestyle. The metaphor of the marathon runner is perfect. Those who run focused exclusively on the finish line, neglecting training, proper nutrition and the body's signals, risk injury before they even reach the halfway point. And the worst part is that, as well as jeopardizing this specific race, it can also jeopardize all future races to come.

This book brings together relevant discussions, insights and case studies, with the aim of offering a different path. It is not a manual with fixed rules or ready-made solutions, because every organization is unique. It is, above all, an invitation to sincere reflection and strategic action. Inspired by the metaphor of the marathon, in which the stages of Preparation, Start, Course and Finish symbolize the different levels of cultural maturity, and by the perseverance of athletes such as John Stephen Akhwari, who showed us the strength of purpose even in the face of pain, the strength of purpose even in the face of pain, let's explore together how to build a culture where safety is, in fact, a value that is lived, practiced and shared every day, in every decision.

Our premise is to promote an urgent change of perspective, abandoning the idea that safety only means the absence of work accidents. This view, however well-intentioned, can lead to complacency, under-reporting and a false sense of control. In its place, we propose an approach in which safety is the active building of capabilities, i.e. an organization's ability to anticipate risks, deal with variability, adapt to the unexpected and, above all, learn to err safely. Safety does not depend on luck but is the result of resilience built up carefully over time.

We will discuss how leadership, at all levels, is the essential engine that drives cultural transformation. We'll see that leading goes beyond setting rules or demanding results. it is about inspiring by example, communicating the "why" of safety in a way that connects with people's deepest values, cultivating an environment of psychological safety where mistakes can be seen as learning and building trust based on coherence between discourse and practice. Leadership sets the tone, defines priorities and demonstrates, in day-to-day decisions, whether safety is really a value or just a priority subject to trade-offs.

We will also explore how rituals (such as meetings, integrations, investigations), symbols (recognitions, visual communication, physical environment) and everyday practices (from issuing work permits to the way we conduct audits or offer feedback) shape collective beliefs and behaviors, forming the fabric of culture. Practices (from issuing work permits to conducting an audit or offering feedback) shape collective beliefs and behaviors, forming the fabric of culture. And, just as Sir Edmund Hillary needed a careful expedition before reaching the top of Everest, we will argue that an in-depth cultural diagnosis is essential to map out the organization's terrain, understand perceptions, behaviors and the way it works. Its terrain, understand existing perceptions, challenges and strengths, and plan the transformation journey effectively, wisely choosing the "right side of the mountain" to start climbing.

Throughout these pages, you will find not only theoretical concepts and practical experiences from around the world, but also applicable tools, real examples from different sectors (adapted to preserve confidentiality, without losing the essence of the lessons), checklists and questions that invite reflection. Real examples from various sectors (adapted to preserve confidentiality without losing the essence of the lessons), checklists and questions that invite reflection, with the aim of helping you to:

Understanding the stages of maturity: Clearly recognizing the different levels of the cultural journey, which include the reactive, dependent, independent and interdependent phases, identifying with absolute honesty which one your organization or team is really in.

Diagnosing your culture: Going beyond superficial indicators to map forces, invisible barriers and assumptions that directly influence safety behaviors in your context.

Engaging leadership strategically: Creating effective ways to gain the genuine commitment of leadership at all levels, from direct supervision to the C-Level and the Board, speaking the language of the business and connecting safety to the delivery of value.

Strengthening the human pillar: Investing wisely in skills development, promoting authentic engagement and cultivating an environment of active care, where psychological safety allows people to feel safe to speak, listen and learn.

Optimize processes: Review and simplify systems and procedures, making them closer to the real work, as it actually happens (Work-as-Done), and more effective in building solid defenses and promoting resilience.

Measure what matters: Implement Proactive Indicators that reveal the health of the culture and the strength of preventive capabilities, going beyond counting failures and looking at what sustains integrity before something happens.

Continuous learning and improvement: Create living mechanisms to learn from mistakes, incidents and, above all, from what works every day, transforming these lessons into concrete improvements.

Sustaining the transformation: Integrating safety into the strategic essence of the organization, into its deepest values, ensuring that cultural change is not just a program with a beginning and an end, but part of the way of being.

Advice for leaders: Cultural transformation starts with you. At the end of each day, ask yourself which of your decisions and attitudes have reinforced safety as a non-negotiable value, and not just as a priority that can be replaced.

This is not a read to simply follow, but a real call to action. Building a solid and resilient Safety Culture is a demanding journey that requires preparation, constancy and courage. Like a marathon, it is not won by the initial push, but by the daily commitment, the firm step and the conscious choice to keep moving forward. Over the next few pages, we'll reflect on how this journey is inevitable for anyone who wants not only to stay on track, but to thrive consistently in a world that keeps changing.

Our invitation is clear. Start or intensify this journey with us. Transform safety from seeing it as an obligation to be fulfilled to seeing it as a real advantage and, above all, as a concrete expression of care for what is essential in every organization - people.

1

DECIPHERING SAFETY CULTURE

BEYOND THE SURFACE

What is Safety Culture? More than catchphrases or numbers in reports, Safety Culture is the way things happen, not what you believe, or hope will happen. It is what is rooted in everyday life, in decisions, gestures and choices, even when no one is looking. It is a shared set of beliefs, values, norms and behaviors that shape how people perceive and deal with safety daily.

It is not something imposed but built. It is born out of daily interactions, genuine listening and, above all, what the leadership chooses to value consistently and coherently.

THE LAYERS OF CULTURE (THE ICEBERG):

To understand culture, we need to look beyond the obvious:

1. **Visible artifacts:** The apparent part of the iceberg. Policies, PPE, reports. Important elements, but they don't define the culture on their own.

2. **Declared values:** What the organization claims to value, such as "it is our greatest value". The distance between discourse and practice needs to be carefully observed.

3. **Unconscious basic assumptions:** The invisible and most influential base of the iceberg. These are deep-seated beliefs, often not verbalized, which shape everyday behavior, such as "Production comes first" or "Here, we look after each other".

4. A strong Safety Culture is one in which these three layers are coherently connected, revealing consistency between what is said, what is done and what is really believed.

Cultural transformation is a journey, not an event. The marathon metaphor helps us visualize this evolution in four phases:

Level 1: Reactive ("Non-Preparedness"): Safety is still seen as an obligation or a reaction to incidents that have already occurred. Leadership remains distant, and a mentality prevails that passively accepts the idea that work accidents just happen.

Level 2: Dependent: The focus remains on rules and supervision, with safe behavior dependent solely on external control. A single phrase echoes repeatedly, reinforcing this restrictive view of "follow the rules".

Level 3: Independent ("Journey"): People begin to take responsibility for themselves, and the discourse clearly changes, echoing a new awareness of personal care, translated into the phrase "I take care of myself".

Level 4: Interdependent ("Arrival and Beyond"): Safety becomes a value shared by all, manifesting itself in genuine cooperation and constant mutual care. The culture is then clearly expressed through the phrase "we take care of each other".

Progressively and consciously moving the organization towards the interdependent level, where safety is experienced as a collective value, care is mutual and each person recognizes themselves as an essential part of mutual and each person recognizes themselves as an essential part of building a safer, more sustainable and humane environment, must be the central objective of this journey.

The need for an in-depth diagnosis (the Everest expedition):

Trying to change the culture without first understanding is like trying to climb Everest without a map. An in-depth diagnosis, combining quantitative and qualitative methods such as questionnaires, interviews and observations, is essential:

- **Mapping the terrain:** Understanding the beliefs and assumptions that really guide behavior.
- **Identify barriers:** Recognize the obstacles that hinder progress and the facilitators that can drive change.
- **Clearly assessing current maturity:** Recognizing where we are in the marathon.
- **Define a strategic and personalized route:** Plan cultural transformation, respecting the pace and reality of the organization.

What, in fact, do we mean when we talk about "Safety Culture"? This term has become a constant presence in corporate vocabulary, appearing in reports, policies and speeches. Even so, its understanding often remains on the surface, reduced to catchphrases or counting days without work accidents. But to start a real journey of transformation, you must go further. You need to understand in depth what this culture is made of within your organization.

In essence, organizational culture, and therefore Safety Culture, can be described simply as "the way we do things around here". However, this simplicity carries a powerful complexity. We are talking about a shared set of beliefs, values, unwritten norms, silent assumptions and patterns of behavior that directly influence how people perceive and deal with risk and safety daily.

It is crucial to understand that culture is not imposed by a memorandum, nor is it installed with a new set of rules. It is born organically, in everyday life. It emerges from conversations in the hallway, from decisions that are made and, just as heavily, from those that are avoided. It is shaped by the stories that are repeated, by the examples that gain strength, by the heroes and villains that are built up within the organization itself, whether they are named or silent.

Above all, culture reflects what is consistently valued, recognized and rewarded by the leadership and the system. If, in practice, what is exalted is fast production, even at the cost of safety, this is what will become true for people, even if the posters say otherwise.

THE LAYERS OF CULTURE: FROM THE VISIBLE TO THE DEEP (THE ICEBERG MODEL)

To understand Safety Culture, we can resort to the iceberg metaphor, popularized by Edgar Schein, or imagine it as an onion, with layers that go from the most visible to the deepest:

Visible artifacts (the tip of the iceberg): This is the most superficial layer, what we can easily see, hear and feel in an organization. It includes tangible elements such as:

- Written policies and procedures (the safety manual on the shelf).
- Use (or non-use) of Personal Protective Equipment (PPE).
- Safety signs and warnings.
- Incident reports and statistics published.
- Formal organizational structure, logos and slogans (e.g. "Safety first!").
- The physical design of the workplace.

Important: Although they are the easiest to observe and change, these artifacts alone do not define the culture. An organization can have impeccable procedures on paper and display lots of safety posters but still carry a deeply reactive and dysfunctional culture in practice.

Declared values (just below the waterline): Here we find the values, strategies, goals and philosophies that the organization publicly claims to have regarding safety. They are the justifications that the leadership offers for its actions and decisions. Examples include:

- Mission and vision statements that mention safety.
- Corporate goals such as "Zero Work Accidents".
- Principles of conduct or codes of ethics.
- Speeches by top management emphasizing the importance of safety.

The Challenge of the Gap: The critical point here is the potential (and frequent) discrepancy between what is declared and what is practiced and valued daily. When the actions of the leadership contradict the declared values, such as pushing for production even in the face of an obvious risk, credibility is lost, and it is the real culture that is revealed.

Unconscious Basic Assumptions (The Submerged Base of the Iceberg): This is the deepest, most powerful and often invisible layer of culture. They are the beliefs, perceptions and feelings that are taken for granted, the unspoken truths about how the world works and how people are expected to behave within the organization. These assumptions are mostly unconscious and difficult to name, but they are what actually drive day-to-day behavior. In relation to safety, they can include:

Negative assumptions: "Work accidents are inevitable in this line of work", "Production always comes before safety when the pressure is on", "Reporting errors or concerns is looking for trouble or being frowned upon", "Rules are for novices, we know how to do the job for real", "Safety is the responsibility of the safety department".

Positive assumptions: "Taking care of each other is part of our job", "There is always time to do the job safely", "Mistakes are learning opportunities, not punishment", "Anyone can (and should) stop unsafe work", "Leadership really cares about our well-being". "Errors are learning opportunities, not punishment", "Anyone can (and should) stop unsafe work", "Leadership really cares about our well-being".

Real Engine: It is these basic assumptions that ultimately determine whether the declared artifacts and values are genuine or just window dressing.

Alignment is power: A strong and mature Safety Culture is one where there is consistent alignment between these three layers. The visible artifacts (such as clear procedures and appropriate PPE) must support and reflect the stated values (safety as a genuine priority), and both must be an authentic reflection of the shared basic assumptions (belief in mutual care, active prevention and continuous learning). When there is misalignment, the culture is weak, confused and prone to failure.

A strong and mature Safety Culture is one in which there is consistent alignment between its three layers. The visible artifacts, such as procedures and PPE, support and reflect the stated values, especially safety as a core value, both of which are genuine expressions of shared basic assumptions, such as the belief in mutual care, constant prevention and continuous learning.

THE MARATHON METAPHOR: LEVELS OF CULTURAL MATURITY

Transforming Safety Culture is not a 100-meter sprint; it is, in fact, a marathon. It is a long journey that requires careful preparation, an adjustable strategy, the stamina to overcome challenges and, above all, the sensitivity to understand the terrain. The beauty of this metaphor lies precisely in its ability to clearly represent the different levels of cultural maturity that an organization can, and needs to, go through.

LEVEL 1: REACTIVE (THE "NON-PREPARATION" OR "DECIDING TO RUN THE DAY BEFORE"):

What it looks like: Safety is seen as a burden, a minimum legal requirement or something that only gains attention after something goes wrong. The organization works in "putting out fires" mode, reacting to problems as they arise.

Mentality: "Work accidents are inevitable", "This has never happened before", "It was the worker's fault", "Safety slows down production".

Characteristics: Metrics are almost always reactive, such as work accident rates and days lost. Responsibility for safety is pushed down to a specific department. Leadership is absent, inconstant or silent. There is little or no investment in preventive actions.

The analogy: It is like the runner who, on the eve of the race, decides to take on a marathon without having trained. They just rely on luck or the willpower of the moment. What is the likely outcome? Injury, giving up or a performance full of suffering.

LEVEL 2: DEPENDENT (THE "START" - FOCUS ON RULES AND SUPERVISION):

What it looks like: The organization begins to recognize the importance of safety, often motivated by regulatory requirements or striking experiences from the past, and begins to structure basic rules, procedures and training. Safety is driven from the top down, by management and supervision.

Mentality: "Safety is following the rules", "If everyone does what we say, we'll be safe", "The supervisor is responsible for ensuring safety".

Characteristics: The focus is on ensuring that people comply with what has been defined. There is a strong dependence on external control, discipline and surveillance. Work accident rates may show some improvement, but the culture is still fragile, sustained by the constant presence of leadership and pressure from the system.

The Analogy: It is like the runner who has started training and is following a basic plan given to him by his coach. He runs, but still depends on the instructions, the pace imposed by the group and external motivation. He hasn't internalized the discipline yet. He runs because he's been told to, not because he understands why.

LEVEL 3: INDEPENDENT (THE "PATH" - FOCUS ON INDIVIDUAL RESPONSIBILITY):

What it looks like People begin to understand, in a deeper way, the importance of safety and take individual responsibility for their choices. There is more awareness of risks and procedures, and workers start to take care of themselves more consistently.

Mentality: "I am responsible for my safety", "I know the risks and how to protect myself", "I follow the procedures because I understand their importance".

Characteristics: The focus is on individual behavior, strengthening safe habits and self-discipline. PPE is used properly, procedures are followed with understanding, and there is a more active attitude to identifying risks. Work accident rates continue to fall.

The Analogy: It is like the runner who already has autonomy, knows their own body, knows the pace they can sustain, takes care of hydration, nutrition and follows the training plan with discipline. They run with awareness, guided by personal commitment.

LEVEL 4: INTERDEPENDENT (THE "ARRIVAL" AND BEYOND - FOCUS ON MUTUAL CARE AND THE SYSTEM):

What it looks like: This is the most mature and desired level, the finishing point of the cultural marathon. Safety ceases to be just a procedure and becomes a shared value, present in the way the organization thinks, decides and acts. People not only take care of themselves, but also of each other, naturally and intentionally.

Mentality: "We are responsible for each other's safety", "Safety is a team effort", "We can always improve", "Reporting errors and concerns make us stronger".

Characteristics: There is a collective sense of belonging and responsibility. People intervene when they perceive risks, share learning, support each other, seek solutions together and strengthen the system. Care is active and mutual. There is a focus on prevention, continuous improvement and building resilience. Leadership acts as a facilitator, promotes psychological safety and creates spaces for listening and exchange. Work accident rates are low and, what's more, sustainable.

The Analogy: This is the experienced runner who crosses the finish line with consistency, but who also encourages others, shares learnings, helps those in need and contributes to the organization of the race. He understands that success is collective and runs for something bigger than himself.

Understanding the Dynamics: It is important to recognize that these levels are not fixed and perfectly separate stages. In the same organization, you can find areas, teams or departments at different points in their journey. While maintenance may operate interdependently, a new production line may still be at the dependent level. The transformation happens little by little, and the aim is to lead the organization to move towards interdependence, until care is, in fact, a value that is lived everywhere.

THE NEED FOR A DEEP DIAGNOSIS: THE EVEREST RECONNAISSANCE EXPEDITION

Before embarking on any serious journey of cultural transformation, it is absolutely essential to know where we are. Trying to implement safety initiatives, however well-intentioned, without an accurate diagnosis of the current culture is like trying to climb Mount Everest blindly, without a map of the terrain, without knowing the weather ahead and without sincerely assessing the strengths and limitations of the team. The most likely outcome? Losing our way, running out of resources or, worse, suffering a work accident that could have been avoided.

Let's remember the story of Sir Edmund Hillary. He didn't reach the top of Everest on his first attempt. His achievement, in 1953, was possible because, the previous year, a reconnaissance expedition led by Eric Shipton mapped out the main challenges. The team crossed the Khumbu icefall, identified the feasible route via the West Valley and the Lhotse face, and proved that the summit was accessible from the south side. Without this careful look at the terrain, the 1953 victory might not have happened.

IN THE SAME WAY, A DEEP AND HONEST CULTURAL DIAGNOSIS IN SAFETY ALLOWS US TO:

Mapping the cultural terrain: Going beyond assumptions and genuinely understanding the perceptions, beliefs and assumptions that exist in relation to security in the different areas, hierarchical levels and working groups of the organization. Safety in the different areas, hierarchical levels and work groups of the organization. What do people really think? What do they feel? What is said and what remains silent?

Find out what is really holding back the culture, such as a lack of trust in leadership, bureaucratic processes that encourage shortcuts, fear of retaliation or conflicts between areas. And, at the same time, identify what can serve as a starting point, such as cohesive teams, safe practices that already exist, committed leaders or positive responses in times of crisis.

Assessing real maturity: Honestly understanding what stage of the marathon the organization, or specific areas of it, really are at. We need to go beyond the traditional work accident metrics and examine in depth whether we are, in fact, at the independent level or whether strong traits of dependency or even reactivity still persist.

Define the strategic route: Based on the diagnosis, build a strategic plan that is realistic, personalized and focused on the areas with the greatest potential for impact. It means wisely choosing the right side of the mountain to start climbing, directing energy and resources to where there is the greatest possibility of traction and consistent progress.

HOW TO DIAGNOSE? BEYOND CLIMATE SURVEYS

This in-depth diagnosis cannot be made through generic organizational climate surveys or simple analysis of work accident data (which only show the past). It requires a multifaceted and triangulated approach, combining:

QUANTITATIVE METHODS:

Directed and online questionnaire: Application of surveys, both online and face-to-face, to obtain a broad view of workers' perceptions of the elements of Safety Culture.

QUALITATIVE METHODS (ESSENTIAL!):

Site visit with behavior observation: This involves going to the workplace and making direct observations of workers' behavior, identifying safe practices and deviations.

Interviews with top and middle management: Conducted to capture perceptions, attitudes and the level of involvement of the leadership with safety, usually lasting between 40 minutes and 1 hour.

Documentation analysis: Evaluation of documents that represent the management of safety culture, such as policies, procedures, training records and internal communications.

Focus groups: Discussion groups with middle management, operational front line and third parties, aimed at understanding perceptions, experiences and points for improvement.

It is crucial to listen to the voices on the front line, to understand their lived experiences, their frustrations, their ideas and their "little ways" (which are often signs of systemic problems). We need to observe how critical decisions are made in practice, especially when there are conflicts between production and safety, and how leadership (at all levels) actually behaves on a day-to-day basis - their actions speak much louder than their speeches.

Practical diagnostic advice: Don't rely exclusively on a single source of data. Combine questionnaires, open conversations and direct observations to achieve a broader, more concrete and sensitive view of your safety culture. Only with this deep understanding, grasping the visible and invisible layers of the culture, its strengths and weaknesses, its coherences and misalignments, can you draw up a plan that is both viable and transformative. Now, with the map in hand, we move on to the strategies and tools that will support this construction. It all starts with an essential change of perspective, because safety is not just the absence of work accidents, but the living, intentional and constant presence of robust capacities to deal with the risks that permeate our reality.

QUESTIONS FOR REFLECTION

1. What are the main visible artifacts related to safety? What are the most common stated values when it comes to safety? Can you identify any gaps between these stated values and the practices you actually observe daily? What basic assumptions (positive or negative) do you believe drive people's behavior in relation to safety?

2. At which level would you place most of your organization today? (Reactive, dependent, independent or interdependent). Are there areas or teams that are clearly operating at different levels? What practical evidence supports this assessment? (For example: observed behaviors, decisions made, leadership attitudes, active participation, reports from the front line, etc.)

3. Has your organization carried out any kind of formal Safety Culture diagnosis? If so, what methods were used? Have the results actually been used to direct concrete actions? If not, what do you think would be the greatest benefit of carrying out an in-depth diagnosis at this time?

4. Assess the degree of alignment between the three layers of culture in your organization: visible artifacts, stated values and basic assumptions. Where do you see the biggest misalignments? How do these misalignments impact day-to-day safety, decisions, behaviors and the way people feel and protect themselves at work?

2

THE CULTURAL TRANSFORMATION MARATHON

Starting a journey of transformation into Safety Culture is undoubtedly comparable to preparing for a marathon. We have a clear starting point and set ourselves an ambitious goal, which is to cross the finish line towards a truly safe working environment in which people can flourish. However, the road ahead is long, full of steep climbs, unexpected descents and stretches that test not only our physical strength, but also the depth of our conviction.

This journey requires much more than an initial burst of enthusiasm. It requires clear purpose, constant perseverance, strategy with intention and attentive listening to the cultural terrain we are treading, with its nuances, histories and complexities.

How many of us have been there? We start off full of energy, perhaps driven by a milestone event, inspiring benchmarking or even a thoroughly prepared plan. However, as time goes on, the results don't come as quickly as we'd like. Obstacles crop up along the way, such as silent resistance, competing priorities and unexpected budget cuts. Energy dissipates, skepticism arises and suddenly we find ourselves in the middle of the journey, discouraged, questioning everything or, even worse, unclear about how to proceed.

It is precisely in these moments that the essence of cultural transformation reveals itself. Because it is not a short race, decided in the excitement of the start. It is an ultramarathon. And it is won with firm and consistent steps, every day.

The inspiring story of Tanzanian marathon runner John Stephen Akhwari at the 1968 Mexico City Olympics serves as a powerful and visceral metaphor for this journey. Akhwari traveled thousands of kilometers, proudly representing his country, and prepared hard for the race. However, right at the start of the race, amid the confusion and jostling at the start, he suffered a bad fall, seriously injuring his knee and shoulder. The pain was excruciating. He had the perfect excuse, perhaps even a doctor's recommendation, to give up. No one would blame him.

But Akhwari hadn't traveled thousands of kilometers just to start the race. His purpose was greater, deeper. With his leg bandaged in an improvised way by a doctor who tried to convince him to stop, limping visibly and fighting the pain with every step, he continued.

He arrived at the Olympic stadium more than an hour after the winner, when the awards ceremony had already finished and most of the spectators had left, leaving the stands almost empty. The sun was setting.

Asked by a stunned journalist why he hadn't given up in such adverse circumstances, his simple but profound answer echoed through time, capturing the true spirit of perseverance and purpose:

"My country didn't send me 5,000 miles to start the race, but they sent me 5,000 miles to finish it."

This mentality, Akhwari's mentality, is essential in transforming Safety Culture. It is not just about launching pretty programs, applying trendy tools or complying with a mandatory training schedule. It is about having clarity of purpose. Why are we really doing this? What is our "finish line"?

And, more than that, it is about sticking with it even when the results don't come at the expected speed, when the obstacles seem bigger, when resistance appears subtly or when criticism comes for not delivering "quick wins". When the obstacles seem bigger than our breath, when resistance appears subtly or when we are criticized for not delivering "quick wins".

The real victory in a cultural marathon is not just in setting off with enthusiasm. It is in following through to the end. It lies in building something that makes sense, that transforms, that remains. It is in honoring, every day, the purpose that puts us on this journey.

THE DANGERS OF LAUNCHING: THE TRAP OF IGNORED CONTEXT AND FATAL COPY-PASTE

Akhwari's premature fall from grace serves as a clear warning about one of the most frequent and damaging mistakes in safety management and so many other change initiatives. It is the lack of critical attention to the specific context of each organization.

In the rush to improve indicators or adopt the market's "best practices", many companies end up giving in to the temptation to import ready-made solutions.

Whether it is a sophisticated risk analysis form used by a large multinational, a Behavioral Observation program that has been successful in another sector or a promising technological tool for managing work permits, it can all seem seductive.

However, this approach, which we could call "cultural copy-paste", ignores an essential truth. Each organization is a unique ecosystem, carrying its own history and a pre-existing culture full of deep assumptions, informal power relations, specific operational pressures, genuine talents, concrete limitations and challenges that don't always appear in official reports. It is a unique ecosystem that carries its own history and a pre-existing culture full of deep assumptions, informal power relations, specific operational pressures, genuine talents, concrete limitations and challenges that don't always appear in official reports.

Copying and pasting practices, no matter how inspiring they may be on paper, without careful adaptation and without listening deeply to the reality of the situation, are like trying to run a marathon in someone else's shoes, on unfamiliar terrain and with a generic map. The fall, the frustration and the waste of energy and resources are not long in coming.

Practical example: Think of a company in the construction sector that has tried to rigidly implement a complex, robust electronic work permit system, copied from an oil refinery where the processes are more standardized and the team is mostly made up of specialized technical staff.

What happened? The bureaucracy created by the system simply didn't fit in with the fast-paced and variable dynamics of construction sites. Under pressure to deliver on time, workers and supervisors began to find parallel paths, the famous workarounds. Forms began to be filled in just to meet requirements, without proper analysis, and many teams reverted to old, informal practices.

The intention was good. The tool was modern. But the lack of adaptation to the context turned the initiative into something not only ineffective, but also dangerous, because it generated a false sense of control.

Studies and field experience have repeatedly confirmed that most cultural transformation initiatives fail precisely because they ignore the real context. Before implementing any solution, no matter how promising it may seem, it is crucial:

1. **Carry out an in-depth diagnosis:** As discussed in Chapter 1, it is essential to understand the current culture, basic assumptions and perceptions of the front line. Without this insight, any intervention risks being superficial.

2. **Active listening:** Sitting down with those who do it, with those who live the process every day. Listen to their difficulties, their ideas, their adaptations. it is in this listening space that the keys to real solutions lie.

3. **Analyze the actual processes ("work-as-done"):** Observe how work actually happens, not just how it is described in procedures. Practical experience often reveals gaps that paper does not.

4. **Identify specific barriers and facilitators:** Understand what, in the organization's reality, hinders or strengthens safety. What stands in the way? What drives it? it is based on these answers that a possible and real transformation can be built.

Only with this contextualized understanding can we begin to design or adapt solutions that have a real chance of success and sustainability.

Practical advice: Before adopting an external "best practice", ask yourself: "What are the fundamental differences between the context where this practice worked and our current context? How do we need to adapt it so that it really adds value here?"

NAVIGATING THE MIDDLE OF THE RACE: PERSEVERANCE, LEARNING AND THE TRAPS OF ILLUSION

If the start is dangerous due to the lack of context, the middle of the cultural marathon brings its own challenges, and they may be even more treacherous. The journey of Akhwari, who continued to run even though he was injured, leaves us with a valuable lesson about the importance of informed perseverance in the face of adversity, which will arise sooner or later.

Even with a well-made diagnosis and a carefully designed strategic plan, transformation almost never follows a straight, continuous, upward line. The "middle of the road" is usually marked by:

- **Slow or stagnant results:** The improvements perceived at the beginning may lose momentum, and the indicators may even worsen temporarily, generating frustration and doubts about the way forward.

- **Increased resistance:** As changes take shape and touch on actual practice, resistance, whether subtle or outspoken, tends to intensify.
- **Change fatigue:** The initial enthusiasm **wears off**. The continuous effort begins to take its toll, and even the leadership can show signs of fatigue.
- **Adverse events:** Work accidents or relevant incidents can arise along the way, shaking confidence in the process and testing, in a real way, commitment to the journey.

In these critical moments, the temptation to abandon the plan, doubt the choices made, look for culprits or simply go back to old practices is immense. This is where clarity of purpose and organizational resilience are tested to the limit.

THE RIGHT ANSWER: INFORMED PERSISTENCE, NOT BLIND STUBBORNNESS

If we have a solid diagnosis on which to base our plan, a coherent strategy and a governance structure to monitor progress, the response to obstacles should not be hasty abandonment, but informed persistence. This means:

1. **Analyze, don't react:** Investigate in depth why results are not appearing or why resistance has increased. Avoid hasty decisions and impulsive reactions.
2. **Learning from challenges:** View obstacles, and even failures, as valuable opportunities to learn about the system, relationships and culture.
3. **Adjust course, don't abandon ship:** correct course, when necessary, refine strategies, intensify dialog and listening, but without losing sight of the greater purpose and the principles that underpin the journey.

BEWARE OF THE ORGANIZATIONAL DISEASES OF COMPLACENCY:

The anxious search for immediate results, or the pressure to show progress at any cost, can lead us to fall prey to two dangerous organizational "diseases", which mask reality and delay genuine transformation:

The illusion of conformity: Believing that everything is going well and that the culture is evolving just because superficial Reactive Indicators, such as the lost time accident rate, show a momentary improvement or because previously announced audits have had good results, is a limited view. This perspective ignores what is really going on below the surface, such as the latent risks and the real behaviors present in the organization's daily life.

The ignorance of optimism (or "Wishful Thinking"): Expecting positive results from actions that, in practice, are empty. Initiatives that don't engage with the reality of the job, that don't address the root causes identified in the diagnosis and that exist only to "check the box". It is implemented out of obligation, not conviction. It is expecting transformation where there is no substance.

FINDING THE RIGHT CADENCE:

Just as Akhwari, despite limping, found a rhythm that allowed him to continue and finish the race, organizations also need to find their own sustainable cadence of transformation. This can't be built on the fly; it rests on four essential pillars:

Coherence: Fidelity to the principles and values defined for the transformation. Actions must reflect discourse, because where there is contradiction, there is a loss of trust.

Consistency: Uniform application of the new practices and expectations throughout the organization (or target area), without exceptions that weaken the credibility of the process.

Constancy: Maintaining effort and focus over time, even when the results don't come at the desired speed. Cultural transformation requires time, patience and presence.

Courage: Willingness to face the difficult truths that a diagnosis reveals. To make decisions that won't always be popular but are necessary. Courage to stick to a commitment, even when the pressure is on.

Navigating the middle of the race requires discipline, continuous learning and leadership that is visibly committed to staying on course, adjusting the sails, when necessary, but without losing sight of the destination.

CROSSING THE FINISH LINE: REDEFINING VICTORY - BUILDING ROBUST CAPABILITIES

Akhwari's epic journey, which culminated in his lonely and painful arrival at the stadium, offers us perhaps the most profound and decisive lesson of this chapter.

It is necessary to redefine the meaning of crossing the finish line when it comes to safety culture.

For a long time, especially in the more reactive and dependent stages of maturity, the idea of victory in safety was associated exclusively with the absence of work accidents, measured by Reactive Indicators such as frequency and severity rates.

Although reducing damage is an essential and non-negotiable goal, Akhwari's journey teaches us that focusing only on the absence of failures and, in his case, simply not giving up is too narrow-minded.

The real victory, the one that gives meaning to every step of this long and demanding marathon of cultural transformation, lies in something greater. It is not just about avoiding the negative, but actively and intentionally building the positive. It is about developing solid organizational capabilities that allow us not only to face the natural challenges of the job, but to learn from them, evolve from them and thrive in spite of them. That is the finish line that is worth crossing.

WHAT ARE ROBUST CAPABILITIES? BEYOND ZERO WORK ACCIDENTS

Going beyond focusing exclusively on the absence of work accidents means focusing on the presence and strengthening of essential capacities, such as:

Resilience: The ability of systems, people, processes and technology to anticipate, monitor, respond to and learn from unexpected events. It is adapting to keep operations safe even under pressure or in the face of failure. It is not just preventing something from going wrong but ensuring that things continue to work even when something fails.

Anticipation: The ability to proactively identify emerging risks and latent conditions in the system before they become incidents. It involves curiosity, active listening, constant questioning and attention to weak signals.

Continuous learning: The ability to learn deeply from events, be they failures or successes, sharing these learnings and turning them into real changes. This is only possible in a culture with psychological safety, where people feel safe talking about mistakes, vulnerabilities and challenges.

Adaptation: The ability to adjust strategies, processes and behaviors in the face of changes in the environment, without losing focus on safety as a core value. It is about maintaining integrity even in uncertain scenarios.

Practical example: An airline that, in addition to investigating work accidents (reactive approach), invests intensively in CRM (Crew Resource Management) training to improve communication and decision-making in complex situations (resilience), that uses voluntary reporting systems and flight data analysis to identify trends before they become problems (anticipation), and that promotes regular debriefing sessions after challenging flights.it uses voluntary reporting systems and flight data analysis to identify trends before they become problems (anticipation), and promotes regular debriefing sessions after challenging flights, ensuring that lessons learned are shared and applied (continuous learning).

MEASURING WHAT MATTERS:

This change in focus requires going beyond "X days without a work accident" plaque, ISO certification seals and purely Reactive Indicators. We need to develop and consistently monitor proactive and capacity indicators that truly reflect the health of the safety system. These include:

- Quality of incident investigations and implementation of corrective actions.
- Level of leadership engagement in visible and meaningful safety activities.
- Perception of psychological safety and organizational justice among workers.
- Quality and frequency of safety dialogues.
- Effectiveness of change management processes.
- Number and quality of improvements implemented based on workers' suggestions.

THE FINAL CONQUEST: HUMAN AND ORGANIZATIONAL TRANSFORMATION

Completing the cultural transformation marathon, therefore, requires much more than technical preparation or the simple implementation of systems. It requires heart, purpose and humility to recognize that this journey is continuous, that perfection may never be achieved, but that it is precisely the relentless pursuit of improvement that moves and transforms us.

The greatest achievement is not just "overcoming the mountain" of safety or zeroing out Reactive Indicators. It is in transforming ourselves, our leaders and our organization along the way by creating an environment where people not only work safely, but feel safe, valued and able to actively contribute to a more resilient future.

Practical Tip: Change the main question from "How many work accidents have we had?" to "How robust are our defenses and capabilities to deal with the risks we face today and tomorrow?"

QUESTIONS FOR REFLECTION:

1. **Purpose:** What is the "finish line" for your organization in terms of Safety Culture? Is this purpose clear, alive and shared by everyone?

2. **Context vs. Copy-Paste:** Has your organization fallen into the copy-paste trap? What were the practical consequences? What has been done to ensure that upcoming initiatives are, in fact, contextualized?

3. **Perseverance:** When obstacles or negative results arise in the safety journey, what is the reaction of the leadership and the organization? Is there room for analysis and learning, or is the tendency to look for culprits and abandon the plan?

4. **Illusions: Is** your organization at risk of falling into the "illusion of conformity" or the "ignorance of optimism"? What practices or indicators could be masking reality?

5. **Capabilities:** In addition to the absence of work accidents, how does your organization measure or evaluate the actual presence of robust capabilities, such as resilience, anticipation and continuous learning?

We need to go beyond plaques indicating days without work accidents, certification seals and purely Reactive Indicators. We need to measure and strengthen those capabilities that really sustain safety in our daily lives, such as the quality of leadership, the genuine engagement of people, the effectiveness of processes, constant openness to learning and the presence of true organizational justice.

Completing the marathon of cultural transformation requires more than technical preparation. It requires heart, purpose and humility to recognize that this journey is continuous. Because the greatest achievement is not in mastering the "mountain" of safety, but in transforming ourselves and our organization along the way.

3

THE SHARED RACE

LEAVING THE GRANDSTAND TO SUPPORT THE JOURNEY

If the journey of cultural transformation is indeed a marathon, as we saw with Akhwari, it will rarely be a solitary race. Just as an elite marathon runner has a whole support team, including coaches, nutritionists, physiotherapists and even fellow runners who help set the pace, organizations and their leaders, in the quest for a more robust Safety Culture, need something similar. They need strategic alignment, active mutual support and a shared sense of purpose. Without this, it is difficult not only to start, but above all to cross the finish line with integrity and sustainability.

The deeply moving story of British sprinter Derrick Redmond at the Barcelona Olympics in 1992 vividly illustrates this visceral need for human connection, unconditional support and the determined refusal to remain a mere spectator to the suffering of others.

Redmond was one of the big favorites for the gold medal in the 400 meters. After overcoming previous injuries, he was in excellent physical and mental shape. He had already won his heat in the quarterfinals and had started the semi-final with a good chance of reaching the final. However, halfway down the opposite straight, a dry snap echoed around the track. His Achilles tendon had ruptured. Redmond fell immediately, writhing in agony. His face reflected the purest despair as his Olympic dreams, carefully cultivated over years of sacrifice and dedication, seemed to crumble in an instant.

However, in an impressive gesture of courage and determination, like the perseverance shown by Akhwari, Redmond refused to give up. He stood up, ignoring the intense pain, and began to limp slowly, almost hopping on one leg, towards the finish line. He was determined to finish this race, even if he finished last, even if there was no prize left to be won.

What happened next transcended the sporting world, becoming one of the most emotional and symbolic moments in Olympic history. His father, Jim Redmond, who was watching in anguish from the stands, couldn't contain his emotion and protective instinct. Without hesitation, he quickly climbed down, overcame the safety guards, invaded the track and ran to meet Derrick.

He hugged him with tear-filled eyes and whispered in an embarrassed voice that he didn't need to go any further. Derrick, between tears and pain, replied with conviction that yes, he had to finish it. Jim, with firmness and deep love, replied that they would go on together until the end.

Leaning on his father's shoulder, who bore some of the weight and pain, Derrick Redmond crossed the finish line, crying, to the emotional applause of 65,000 people on their feet. He was disqualified for outside help, but at that moment it didn't matter.

Because at that moment, the world saw more than just a race. It saw the strength of true support, the value of the bond that sustains and the beauty of leaving no one behind.

And for those of us who talk about Safety Culture, there's a message that can't be ignored. When someone is injured in the middle of the road, the worst thing we can do is continue to sit in the stands. Support, be there, walk with them. That too is safety, that too is leadership, that too is culture.

THE SILENT CHALLENGE: STRATEGIC MISALIGNMENT AND HIDDEN RESISTANCE

The moving scene of Redmond running with his father's support brings to the fore one of the harshest and often overlooked realities of organizational transformations, which is the challenge posed by strategic misalignment and internal resistance, often silent and invisible to the most hasty of eyes.

Consistent studies and the practical experience of leaders and consultants show a truth that is hard to ignore: a large proportion of change initiatives, including those aimed at Safety Culture, fail not because of a lack of technical planning, resources or the official discourse of the top leadership. What often sabotages these initiatives is precisely the discreet resistance that occurs behind the scenes, coupled with the lack of genuine alignment between the different areas of the organization.

A plan can be structured with excellence, approved by the CEO, disseminated in inspiring meetings, accompanied by carefully designed campaigns and with formally recognized leaders as ambassadors for change. There may be modern tools, validated methodologies and perfectly aligned speeches. However, if peers from other areas, middle managers, operational teams or even people from our closest circle are not truly committed to this journey, walking together in the same direction and with the same purpose, progress will be slow and frustrating at best. At worst, it simply won't happen.

THE HIDDEN FACES OF MISALIGNMENT:

This misalignment rarely appears in an explicit or confrontational way. Often, it operates between the lines, discreet but deeply corrosive. It manifests itself in subtle gestures, veiled statements and attitudes that gradually undermine the consistency and strength of the transformation.

Side conversations and cynicism: Skeptical comments in the corridors, jokes that disqualify the initiative, the famous "peon radio" spreading disinformation or discredit.

Lack of interdepartmental collaboration: Areas that don't talk to each other, don't share what is essential, create bureaucratic barriers or simply don't prioritize what comes from the safety agenda.

Withheld or distorted feedback: Leaders who silence the team's pain or filter information to avoid discomfort, preventing the top from seeing what is really going on.

Conflicting priorities (real or perceived): The old tension between production and safety, where in practice delivery targets continue to override protection needs, despite official discourse.

Superficial engagement (façade compliance): People who are present at meetings, fill in reports and follow protocols, but do everything in automatic mode, without conviction, without any real change in behavior.

COMMON EXAMPLES OF FUNCTIONAL MISALIGNMENT:

Finance: May see safety primarily as a cost center, questioning proactive investments and focusing only on short-term financial returns, without understanding the strategic value of prevention and building resilience.

Human resources: May not consistently integrate safety principles and competencies into recruitment, selection, performance evaluation and leadership development processes.

Internal communication: It can fail to reinforce key safety messages consistently and intentionally or dilute them amid other organizational priorities.

Operations/Production: May perceive safety initiatives as impositions from "above" or "outside", with no connection to operational reality, generating a sense of bureaucracy and hindering the delivery of goals.

How many times have we, as leaders or safety professionals, not felt like Derrick Redmond moments before his father's intervention? Prepared, focused, delivering our best during the race, when suddenly we are hit by some kind of internal "injury" to the organization. It could be an unexpected misalignment or a silent resistance, something we didn't foresee or couldn't deal with effectively, and which ends up paralyzing us.

That is why we need to remember, every day, that transforming safety culture is not the sole responsibility of the ehs area, nor of top management alone.

It requires a joint effort, constant dialogue and true alignment between all areas and levels of the organization. everyone needs to come out of the bleachers.

COMING OUT OF THE BLEACHERS: THE ACTIVE AND ESSENTIAL ROLE OF THE SAFETY PROFESSIONAL (AND ALL LEADERS)

Jim Redmond's decisive move out of the stands and onto the track to actively support his son in his moment of greatest pain offers a powerful and life-changing lesson for safety professionals and, indeed, for leaders at all levels of the organization.

How often do we as safety professionals, perhaps without realizing it, put ourselves in the position of critical spectators? From our "grandstand", be it the EHS room, the corporate office or the position of technical expert, we point out what the leadership should be doing, we bemoan the lack of engagement, we expect an example to be set and changes to happen as planned. The leadership should be doing, we bemoan the lack of engagement, we hope that the example comes ready, that the changes happen as planned, that the operation embraces our proposals without hesitation.

We get frustrated when line staff don't follow procedures, when production prioritizes targets over safety or when supervisors don't communicate in the expected way. We often hear phrases like "leadership needs to ensure that safety happens" or "safety is the line's responsibility". Although these statements contain important truths about accountability and the integration of safety into operations, they can, when unaccompanied by genuine presence and support, lead us to adopt a passive and distant stance, reduced only to collection.

We can't simply transfer responsibility and expect the transformation to happen on its own.

Just as Jim Redmond didn't just shout instructions from the stands, we also need to go down to the track. Be with the leaders, the supervisors, the teams. Feel the terrain with them. Help carry the weight of change. To listen, to support, to walk with them.

It is on the ground, with our feet on the track and our eyes on the goal, that real transformation begins to take shape. Safety is not built from afar; it is built together.

WHAT DOES "GETTING OUT OF THE BLEACHERS" MEAN IN PRACTICE?

This change in attitude takes shape in real attitudes and in a more humane and collaborative mentality:

1. **Building bridges, not walls (relationships):** it is about dedicating real time to understanding what pressures, what challenges and what drives the other areas, such as Operations, Maintenance, Engineering, HR and Finance. It is about realizing that they are not obstacles, but internal partners. This starts with the simple: listening carefully before suggesting any changes, showing genuine interest in each other's work, having a coffee, opening space for an unpretentious conversation. Connection begins where there is listening and empathy. And that sometimes changes everything.

2. **Be a facilitator and coach, not just a supervisor (active support):** Shift the focus from pointing to supporting. Replace the magnifying glass of control with the outstretched hand of guidance. Being with them in the field routines, walking alongside the leaders (not in front or behind them), participating in meetings with an open ear, not just to talk about safety, but to really get to know their day-to-day lives. it is about offering tools, sharing learning and helping these leaders to become stronger. Teaching them, for example, the art of conducting a good safety dialog or investigating an incident with an eye to learning.

3. **Translate and connect, not isolate (facilitate alignment):** Safety needs to be understood by everyone. But to do this, you need to speak the language of each area. Translate safety into what makes sense within each team's universe. Connect safety goals to business goals. Showing that taking care

of people also means taking care of efficiency, quality, reputation and engagement. And that when there is a real dialog between the areas, safety is no longer an isolated effort but part of the whole.

4. **Being present where the action takes place (visible presence):** Giving up the distant, overly administrative or reactive stance. It means going out into the field with your whole heart. Be where the real work happens, be it on the shop floor, on the construction site, in the office or at Daily Safety Talk. But not as someone who arrives to charge, but as someone who arrives to observe, learn, listen and support. With attentive eyes, available ears and hands outstretched to add. Because you can only transform a culture by being together and really being there.

1. **Influencing with credibility, not just authority (being an influencer):** This is when technique stops being just the norm and becomes a bridge. Using knowledge not just to create rules, but to inspire, provoke reflection and generate real change. This requires more than technical mastery. It requires listening, empathy and the courage to negotiate respectfully and present arguments that make sense to those on the front line. Because imposed authority may silence, but it is credibility that opens doors and touches people's hearts.

Practical advice for safety professionals: Take time to simply be present. Observe an activity, talk to an operator, participate in a production meeting without any specific safety script or agenda. Just be. Just listen. Trust is born of presence, and credibility is strengthened when there is partnership.

You can't expect operational leaders to become safety benchmarks overnight, especially if you're not there with them, helping, teaching, challenging with love and running this marathon side by side. Delegating this transformation without offering active presence and real support is like expecting an injured athlete like Derrick Redmond to cross the finish line alone and still get on the podium. It is unfair. It is inhumane. And that is not how you build culture.

THE HUMAN TOUCH: HUMANITY, HUMILITY AND THE LEGACY WE LEAVE BEHIND

The story of Derrick and Jim Redmond, father and son who met in pain, right there in the middle of the Olympic track, goes beyond physical support. It reminds us, with delicacy and strength, that no transformation happens without humanity. Without humility. Without the courage to recognize the pain that inhabits us and the vulnerabilities that surround us.

Looking after a crop is not just a technical task. It is not about applying tools, filling in spreadsheets or following a timetable. Cultural transformation is first and foremost profoundly human. It involves changing mindsets, reframing beliefs, transforming habits and creating new bonds. And this isn't just done with logic or authority. It is done by listening. With presence. With respect.

When we ignore emotions, fears and silence, we put everything at risk. Because there is no real change without connection. And there is no connection where empathy is lacking. That is why transforming culture also means allowing us to be transformed. It means walking together, even when the pain paralyzes. it is reaching out. It is being the father on the track, not to carry, but to ensure that no one must walk alone.

THE POWER OF HUMILITY:

This journey requires, above all, humility:

Admitting that we don't have all the answers: Especially as experts or leaders, we can feel the pressure of having to know everything. But the reality of complex systems is that no one has a monopoly on the truth. Humility allows us to ask genuine questions, listen to different perspectives (especially from the front line) and be open to learning.

Recognizing our own mistakes and biases: We all make mistakes. Humility allows us to admit them openly, learn from them and show others that making mistakes is part of the improvement process, creating an environment of greater psychological safety.

Ask for help when necessary: No one can transform a culture alone. Humility allows us to recognize our limitations and actively seek the support and collaboration of others.

The "know-it-all" safety professional, who imposes solutions without listening, or the proud leader, who never admits fault and blames others for problems, is unlikely to be able to inspire the trust, respect and collaboration needed for meaningful and lasting cultural change. Arrogance creates walls, but humility builds bridges.

BUILDING A LEGACY OF CARE:

By getting out of our box, whatever it may be, and getting actively involved in supporting our colleagues, leaders and teams, with empathy and humility, we build much more than just a more robust Safety Culture. We build a legacy of care, collaboration and human resilience.

Just as the unforgettable image of father and son crossing the finish line together in Barcelona has become an enduring symbol, the actions of genuine support, partnership and humanity that we demonstrate in our own safety "marathon" leave deep and positive marks on the organization. They show, in practice, that we care about each other, that we are on this journey together and that human value is above any short-term goal or result.

FOR REFLECTION:

1. Where is your "grandstand"? At what moments have you only observed, pointed or delegated safety, when you could have reached out, been more present and collaborative?

2. Who needs your support on the "runway" right now? Think of a leader, colleague or team who is going through challenges on the safety journey. What concrete gesture of support can you offer this week?

3. How can you practice humility more in your relationships? Reflect on situations in which you could listen with more presence, ask with more interest or simply recognize that you don't have all the answers and that is okay.

4. What legacy of care, partnership and presence do you want to leave in your organization? What human marks do you want to remain, far beyond numbers and results?

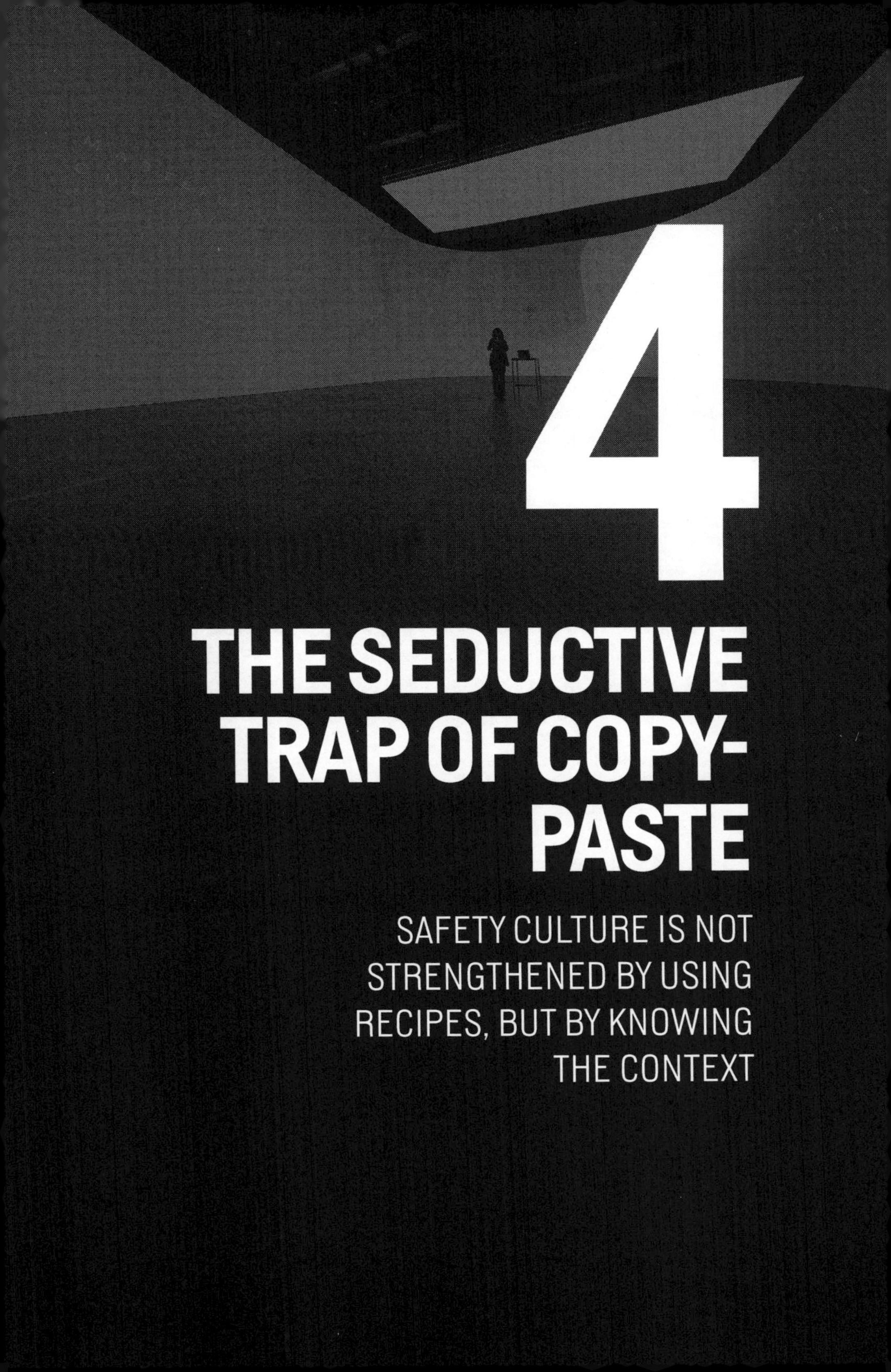

4

THE SEDUCTIVE TRAP OF COPY-PASTE

SAFETY CULTURE IS NOT STRENGTHENED BY USING RECIPES, BUT BY KNOWING THE CONTEXT

In the constant and often anxious search for safety advances and the long-awaited cultural transformation, it is almost instinctive to look outwards. We look for inspiration in the success stories of other organizations, in articles presenting the industry's "best practices", in conferences showcasing award-winning programs and acclaimed methodologies. And, driven by this impulse, we feel a strong temptation to adopt what seems to have worked elsewhere. After all, why reinvent the wheel when someone else already seems to have found the ideal solution?

It is precisely at this point that we encounter one of the most subtle and dangerous traps on this journey, known as the cultural copy-paste trap. This is the silent and mistaken belief that a successful solution in one context can simply be copied and applied in another, with the automatic expectation of obtaining similar results.

Practice and research, however, reveal an important warning. Many studies and experiences show that between 70% and 92% of cultural transformation initiatives fail to achieve their objectives or generate sustainable change. And one of the most common causes is precisely this tendency to copy ready-made models, without looking in depth at the real and unique context of the organization trying to apply them.

Transforming culture is much more than replicating tools. it is about understanding where we are, who we are and what makes sense to our reality. Because culture doesn't care about fads, it responds to the truth of each environment.

WHY DOES THE COPY-PASTE MIRAGE INEVITABLY FAIL?

Imagine trying to use a map of São Paulo to find your way around the streets of Salvador. Or trying to fit one piece of a landscape puzzle into another with a totally abstract image. Sounds strange, doesn't it? But that is exactly what we do when we apply safety solutions copied from other contexts, without thoroughly considering the living reality of our own organization.

Each company is a unique ecosystem, with a history that bears the marks of its pains and achievements. It carries values that are often more present in attitudes than in the pictures on the wall. It has its unspoken norms, the "way things are done here". It has a power network that goes beyond the organizational chart, with influences that are not always based on position, but on trust.

It also has processes that live in two versions: the one drawn up on paper and the one practiced in routine, with its shortcuts, adaptations and improvisations. It also carries specific pressures, tightening targets and deadlines, and a workforce with different profiles, histories and cultures.

What works excellently in a global mining company, with its hierarchical culture, strong union presence and concentrated operational risks, can prove totally ineffective or even harmful in a technological startup with a horizontal structure and fast pace. A successful practice in a referral hospital may not make sense on a construction site or in a transport company with teams spread across the country's roads.

Cultural copy-paste fails because it ignores all this. It tries to simplify what is complex, standardize what is deeply human and adapt without listening. And so, it stumbles over obstacles that are invisible to the eye of the beholder.

1. **Cultural collision (disregards existing culture):** When new practices are imposed without dialog, they often clash with values that are already ingrained. If the culture values autonomy, a change that demands rigid control tends to generate resistance. If there is low trust and openness is expected, the barrier rises even faster. The result is predictable: superficial adherence, cynicism behind the scenes and the old "here comes another initiative that doesn't speak our language" feeling.

2. **The ignored human factor (ignoring people):** All change involves people. And when people are forgotten in the process - their fears, abilities, realities, aspirations - disconnection sets in. A tool may be promising, but if it requires time that doesn't exist or skills that no one has developed, it won't flourish. Policies that punish without welcoming generate fear, silence and under-reporting. And then safety ceases to be a value and becomes a threat.

3. **Clash with operational reality (not adapted to real processes):** There is work on paper and there is work as it happens. When practice is imported without considering this "real world", what is created is not a solution, but an obstacle. Increased bureaucracy, new bottlenecks, team frustration and even additional risks, as improvised shortcuts emerge to survive what has been poorly adapted.

4. **The shadow of the past (underestimating history):** Every organization has a memory. Scars of abandoned projects, unfulfilled promises, speeches disconnected from practice. To ignore this baggage is to ignore the main

indicator of trust: history. And when this history screams "we've seen this before and it didn't work", any new proposal begins with an extra weight of distrust.

5. **The "I didn't invent it" syndrome (lack of "ownership"):** Changes are born fragile when they arrive too ready-made. Without listening, without building together, they don't create belongings. And what doesn't belong doesn't stand up. It becomes "something from safety", "from HR", "from above". It is never "our way". And without this "our way", no practice takes root.

Practical example: A company in the chemical sector, inspired by a successful case in aviation, decided to implement a strict system of checklists for all maintenance tasks, copying the exact model. But the culture there was more informal, based on the experience of senior technicians, and the maintenance processes were quite variable. The result? The checklists were seen as bureaucracy that got in the way of real work. They were filled in superficially, or only after the task had been completed, and the initiative brought no improvement in safety. It only generated frustration, resistance and disconnection from practice.

THE MASTER KEY: THE IMPERATIVE NEED FOR DEEP CONTEXTUAL DIAGNOSIS

If "copy-paste" is the trap, what is the key to avoiding it and building a cultural transformation that really works? The answer lies in a deep and honest contextual diagnosis. Before implementing any practice, program or tool, whether internal or external, it is essential to have a clear understanding of the unique terrain on which we are standing.

This dive goes far beyond analyzing work accident figures or applying generic models of cultural maturity. It requires an attentive, patient and sensitive eye. It means feeling the real pulse of the organization. Like an experienced doctor who doesn't prescribe before listening, observing and investigating, we also need to listen to the voices, stories and experiences that make up the company's routine. Only then will it be possible to build something that makes sense and has roots.

HOW TO CARRY OUT AN EFFECTIVE CONTEXTUAL DIAGNOSIS?

There is no single formula, but a robust diagnosis almost always involves a sensitive, attentive and close look at reality. In general, this involves a combination of approaches that help to listen, observe and understand what happens in practice:

1. Active (and empathetic) listening:

Structured and informal conversations: Talk to people from different areas and hierarchical levels. More than collecting data, it is understanding the viewpoint of those who live the routine. Listening with presence, welcoming perceptions, frustrations, ideas and what happens in "work as it is done", not just as it is on paper.

Focus on open questions: Using questions that invite dialog is key. We can, for example, ask employees how safety works in that environment, what factors contribute to or hinder safe working, what they would change if they had the chance and how safety decisions are made in practice.

Guarantee psychological safety: Create an environment where people feel respected and safe to speak truthfully. Often, guaranteeing anonymity and confidentiality makes all the difference.

2. Observe the actual work (go to the Gemba):

Active presence in the field: Being where the work happens. Look with curiosity, not judgment. Being present, not to supervise, but to learn.

Understand adaptations and pressures: Observe the real working conditions, the resources, the urgencies, the improvisations. Understand why shortcuts are taken and what leads people to adapt what has been prescribed.

Dialogue during observation: Whenever it is safe and possible, talk to the person carrying out the task. Ask respectfully: "What are you doing?", "Why are you doing it like that?" and listen carefully.

3. Analyze existing data (with a critical eye):

In addition to Reactive Indicators: Look at incident reports, near misses, inspections, climate surveys, absenteeism, turnover and production data. But don't stop at the numbers.

Looking for the stories behind the numbers: What patterns are repeated? What does this data reveal about the weaknesses of the system? What root causes are there, hidden behind the statistics?

4. Mapping culture (the visible and the invisible):

Practiced vs. declared values: What really guides day-to-day decisions? Is what you live consistent with what you declare?

Unspoken social norms: What behaviors are tolerated? What phrases are repeated and shape behavior? ("Here we don't stop production for anything", "it is better not to talk about it now...")

Symbols and rituals: Where are resources allocated? What is celebrated? What is ignored? These expressions strongly communicate what is valued.

5. Understand the dynamics of power and communication:

Real decision-making: Who decides in practice? Who influences, even without a formal position? Where is the power that moves choices?

Communication flows: Does information circulate freely? Or does it get lost in filters, fears and noise? The way you communicate says a lot about how much you trust.

These five paths are not ready-made formulas, but windows to see culture more truthfully. And only when we see clearly can we care with intention.

Quick checklist for contextual diagnosis:

- [ ] Did I talk to people on the front line about their safety challenges?
- [ ] Have I observed critical tasks being carried out in the field recently?
- [ ] Have I analyzed incident data looking for systemic patterns, not just immediate causes?
- [ ] Do I understand which unwritten norms influence safe (or unsafe) behavior?
- [ ] Do I know who the real influencers of culture are in my area/organization?
- [ ] Do I understand how production pressures/targets affect day-to-day safety decisions?

Only with this deep and multifaceted understanding of the context, this real reading of the terrain, is it possible to critically assess which external practices can be carefully adapted (and almost never simply copied). What's more, it is in this space of listening and understanding that solutions are born that need to be built internally, in partnership with people, so that they are truly effective, accepted with meaning and sustained with consistency over time.

FROM INSPIRATION TO CONTEXTUALIZED ACTION: CO-BUILDING SOLUTIONS THAT WORK

If in-depth diagnosis is the key to understanding the terrain, co-construction is the way to cross it with safety and truth. Instead of looking for a ready-made "silver bullet" or imposing top-down solutions, what really generates engagement, belonging and sustainable results is building together, with those who live the routine, feel the pressures and will be directly impacted by the changes.

This doesn't mean discarding what comes from outside. On the contrary. Good practices and external references can, and should, serve as a starting point. But they need to pass through the filter of the local context, through active listening and collective experience. Only then will solutions emerge that make sense. And what makes sense, stays.

Principles of contextualized co-construction:

1. **Participatory diagnosis:** From the outset, the diagnosis needs to be collective. Including representatives from different areas and hierarchical levels in the collection and analysis of information, through workshops, conversation circles or focus groups, broadens the view of reality and already begins to generate belonging. In addition, this process helps to identify, in a natural way, who the potential mobilizers of change are.

2. **Collaborative design (solution workshops):** With the challenges mapped out, the next step is to bring together different voices, from operations to leadership, in cross-functional working groups. In these workshops, the focus is not just on discussing ideas, but on consciously adapting external solutions or creating paths together that make sense in practice. Open debate, respect for different realities and valuing each contribution are the basis for viable and sustainable solutions.

Practical advice Tools such as Design Thinking or Value Stream Mapping, adapted specifically for safety, are very useful for conducting these workshops in a structured, light-hearted way and with active listening. They encourage joint creation, focusing on what really matters, which is the reality experienced in the workplace.

3. **Piloting and iterative adaptation:** Before expanding any new practice, start small. Test it in one area or team, intentionally. Define simple metrics, collect constant feedback, observe with active listening what worked, what didn't and what needs to be adjusted. Learn from the process. Refine the solution based on real-life experience and be open to adapting or even backtracking if the pilot shows that it doesn't make sense for that context.

4. **Clear and continuous communication of the "Why":** Change without a clear purpose is lost along the way. It is essential to communicate not only what will be done and how, but above all why. What motivated the change? What pain is it trying to solve? What real benefits does it bring to safety, to the team, to everyday life? Use different channels, translate the language for each audience and keep the communication alive, before, during and after.

5. **Focused training and ongoing support:** Training is essential, but it shouldn't just be treated as a one-off event. It is necessary to offer practical training, specifically geared to the day-to-day reality. Above all, it is essential to maintain active support throughout the routine, ensuring constant presence, offering ongoing guidance and creating spaces for frequent clarifications and adjustments. Changes can only be sustained when people are sure that they are not alone on this journey.

THE POWER OF INNER WISDOM:

Adopting external practice can be a valuable starting point and a legitimate source of inspiration. However, it needs to be seen as an ingredient and not a ready-made recipe. Each organization has its own flavor, rhythm and unique way of operating. The true art of cultural transformation lies precisely in the ability to balance what comes in from the outside with what pulses internally, which is the practical wisdom of people who live day-to-day, feel the challenges of the operation on their skin and hold answers that often don't yet appear in books.

It is from this humble, respectful and conscious integration between external knowledge and internal experience that technically viable and, above all, culturally compatible solutions emerge. Solutions that gain genuine buy-in and have a greater chance of lasting, because they are born from the inside out, by people and for people.

Let's remember the lesson learned by John Stephen Akhwari. He didn't copy any other runner's strategy. He adapted his run to his own physical condition, the altitude, the injury he suffered and the purpose that drove him. He completed his journey precisely because he ran aware of himself and the terrain he was running on. In the Safety Culture marathon, context is not just a detail, but the very ground we walk on. Mapping it carefully and adjusting our strategy to it is not just an option, it is a fundamental condition for crossing the finish line with purpose and authenticity.

FOR REFLECTION:

1. **Recent diagnosis:** When was the last time your organization made an in-depth diagnosis of its safety culture, really listening to the front line and observing the work as it happens?

2. **Imported" initiatives:** Think of a recent safety initiative that was inspired by another company. How much was it adapted to your reality? How much did it involve the people who were going to live it? What were the results?

3. **Voice from the front line:** Are there real and effective ways for ideas and concerns from the operation to reach leadership and influence decisions?

4. **Co-construction in practice:** In which current safety process or challenge can you apply the principles of co-construction participatory diagnosis, collaborative design, piloting to build something more adherent and lasting?

5

THE ESSENCE OF SAFETY

BEYOND ZERO, TOWARDS ROBUST CAPABILITIES

For many years, safety management in organizations was guided almost exclusively by the idea of preventing something from going wrong. Success was measured by the countdown of days without accidents, with the well-known "zero work accidents" as the main goal, complemented by the pursuit of zero injuries and zero deviations. This traditional approach focused its efforts on investigating errors after incidents occurred, trying to understand causes often attributed to human error or procedural failures, with the aim of creating barriers to prevent similar events from happening again.

Undoubtedly, this perspective has brought important advances and still plays an essential role in risk prevention. Learning from what hasn't worked remains valuable. However, when this vision becomes the only one adopted to deal with safety, it reveals its limitations. Taking to the extreme of obsessing over "zero work accidents", it can even generate hidden risks, silence, cover up failures, create fears of exposing errors and ward off critical reflection. Paradoxically, this attitude ends up weakening the very Safety Culture it is intended to strengthen.

THE LIMITATIONS OF THE FAILURE-ONLY VIEW:

Focusing only on the absence of failures means ignoring an essential truth about the complex systems in which we operate. Work almost never happens exactly as planned. There is a natural gap between what is stated in procedures, manuals and training - idealized work - and what actually happens on the front line, i.e. real work.

And why is this different? Because the world we live in is dynamic, full of unforeseen events and continuous change. People must constantly adapt to weather variations, equipment that doesn't always cooperate, shortages of time and resources, productivity pressures and countless little details that no manual can fully anticipate. Most of the time, it is precisely these adaptations that keep everything running. It is the professionals who, using their experience, judgment, creativity and, above all, commitment to safety, overcome daily challenges, ensuring that the workplace is carried out efficiently and safely, even in the face of the imperfections inherent in every system.

Day-to-day operational success is not just about "nothing went wrong". It is almost always sustained by a series of small successes, of decisions adjusted in practice, of careful choices made by people who know the reality in which they operate well. When we only look at failures, when our ruler is based only on work accidents or incidents, we only see the surface, the visible part of the iceberg. We miss the chance to learn from what works, to understand what makes these safe adaptations possible and what we can do to strengthen them. That is where real learning lies.

Even more worrying is when the pursuit of "zero work accidents" stops being an aspiration and starts being treated as an absolute, rigid and unquestionable goal. This excessive demand, often cloaked in good intentions, can end up having the opposite effect to that desired. When the total absence of faults becomes the only acceptable measure of success, and when errors are seen as individual faults to be punished, a culture of fear and silence is silently installed.

People begin to hesitate to report small errors, deviations, near misses or unsafe conditions. This is not due to negligence, but as a form of self-protection, motivated by fear of punishment, fear of damaging the team or compromising the area's indicators. And so what goes unsaid accumulates silently. The numbers appear to be in satisfactory condition, but the risks remain hidden, latent, just waiting for an unfortunate combination of factors to reveal themselves in serious events, just when the defenses, weakened by a lack of learning, fail.

In this scenario, subtle but deeply damaging cultural phenomena emerge. These include the pretense of perfection, where leaders and teams pretend that everything is going well, ignoring warnings and distancing themselves from the daily reality on the front line. Or the paralysis caused by fear, where the fear of making mistakes or jeopardizing the goal of zero work accidents blocks innovation, experimentation and authentic learning. In this way, the Safety Culture is no longer nurtured, and the very goal intended to protect ends up eroding the very thing that is most valuable - trust, the ability to listen and a genuine commitment to care.

THE KEY TURNING POINT: FROM NO MISTAKES TO STRENGTHENING WHAT WORKS

Faced with the limitations of the purely reactive view, a more realistic, powerful perspective is gaining ground, aligned with the way work actually happens. It is about seeing safety not just as the absence of work accidents, but as the active presence of conditions capable of sustaining success, even during the uncertainty, variability and complexity of everyday life.

In this new way of understanding safety, the objective is not restricted to preventing something bad from happening. On the contrary, the aim is to intentionally guarantee reliable defenses, adequate resources, flexible processes and, above all, to cultivate collective intelligence that allows for agile adaptations and safe responses, even when things go wrong.

The focus shifts from simply counting failures to understanding what makes success possible on most days. It means learning from what worked and understanding why it worked. It is about reinforcing what already protects, what already works, what already takes care of.

This view does not diminish the importance of learning from failures. On the contrary, it expands and complements this learning, recognizing that to achieve truly sustainable safety, it is essential to observe and value what keeps the operation quietly steady. It is necessary to cultivate the ability to anticipate risks, respond quickly, adjust wisely and learn from all experiences, both those that bring pain and those that teach lightly.

In the end, it is precisely this daily resilience, experienced by real people in real contexts, that keeps the whole system going. it is this capacity that needs to be valued, nurtured and strengthened if we are to truly protect lives in an authentic and lasting way.

THE FOUR ESSENTIAL CAPACITIES OF ORGANIZATIONAL RESILIENCE:

Building this resilience requires the continuous strengthening of four capacities that intertwine and support safety in real environments, with their daily uncertainties and pressures:

1. **Capacity for Anticipation:** This is the ability to look ahead, to perceive what hasn't happened yet, but could happen. Seeing risks that have not yet materialized, imagining future scenarios and paying attention to subtle signs that something could get out of hand. It goes beyond what's on the risk analysis forms. It involves listening to the intuition of the front line, valuing learnings from other organizations and creating safe spaces where future concerns can be voiced and taken seriously.

2. **Monitoring capacity:** It is about keeping your eyes open to notice what's changing now. it is knowing how to identify when something starts to go off course, whether it is an unsafe condition, a silent pressure or even a seemingly small deviation that can quickly evolve. Monitoring properly involves robust systems, but it depends even more on people who are attentive, present and empowered to speak up. People who understand exactly what to observe and are confident enough to report what they notice, free from fear.

3. **Responsiveness:** Acting with clarity and agility in the face of the unexpected. Having resources ready, trained people, the autonomy to make decisions and, above all, the flexibility to adapt when the plan no longer works. The right answer won't always be in the procedure. Often, it will come from listening, experience and the courage to improvise responsibly when the real scenario requires another course of action.

4. **Ability to learn:** Transforming experiences, good or bad, into a source of growth. Learning not just from what hurts, but from what protects. From what almost happened, from what was avoided, from what worked. This requires an environment that welcomes mistakes as opportunities and treats learning as part of the job. Where knowledge is not stored away, but shared, applied and kept alive in everyday life.

Practical advice: Think of these four skills as muscles of resilience. Just like in the body, it is no good having one strong muscle and the others weak. An organization will only be truly resilient if it strengthens all these points and makes them work in an integrated way, together sustaining safety daily.

FAILING WITH LUCK VS. FAILING WITH SAFETY: THE CRUCIAL DIFFERENCE IN RESILIENCE

The difference between an organization that only seems safe, with few work accidents but sustained by silence or luck, and a truly resilient organization lies in how it deals with the inevitability of error. Because in complex systems, with pressures, variability and so many interconnected parts, failures are going to happen. The question is not whether we will fail, but how prepared we are to react when we do.

Lucky Failure: This is the scenario in which something goes wrong, a detour, a mistake, a barrier that fails, but, by chance, nothing serious happens. It may be that someone notices in time, that a favorable condition helps, that the impact doesn't happen by pure coincidence. Luck protects, but structure does not.

The risk here is silent. Firstly, because the organization may not even realize how close it was to a serious event. And secondly, because as nothing has happened, the opportunity to learn is gone. Nothing is reinforced, no changes are made, and the fragility remains there, waiting for a day when luck won't be around.

The feeling of safety grows, but not because the system is stronger, but because the failure has gone unnoticed. it is the kind of tranquility that soothes on the outside but fragilizes on the inside.

Illustrative example:

Lucky miss: A critical procedure is skipped because of time pressure. This time, nothing happened because the conditions were stable. The failure goes unnoticed, and the organization doesn't investigate the reason for the deviation. The vulnerability remains silent, the system remains exposed, relying on luck to avoid the next event.

Failing safely: This is when, even in the face of an error or deviation, the system can respond with maturity and structure. The capacities to anticipate, monitor, respond and learn come into play. Defenses organized in technological, human and procedural layers act to:

- Detect errors or deviations quickly.
- Contain the impact and prevent it from spreading.
- Reduce the consequences.
- Recover the system with agility and control.

The system is built with the awareness that failures can happen, which is why it has mechanisms in place to withstand them without collapsing. Even without a work accident, failure is recognized. The organization investigates not only what went wrong, but what worked well, and it learns from this. Knowledge becomes adjustment, reinforcement, continuous improvement.

Illustrative example: An operator makes a mistake when configuring a piece of equipment. A cross-checking system detects the inconsistency before the operation starts. Or a sensor fails, but the backup system takes over and trained staff respond quickly, stabilizing the process. The event is reported, analyzed in depth (what caused the error? Why did the defense work? Can we strengthen it?) and the learning is shared with everyone. Because real safety isn't hidden, it is built. Together.

CREATING SAFE SPACES EVEN IN THE MIDST OF FAILURES

Failing with luck is trusting that the worst won't happen. It is crossing your fingers hoping that everything will go well, even in the face of clear signs of risk. Failing with safety, on the other hand, is recognizing that failure is part of the process and therefore needs to be predicted, understood and managed. Instead of relying on luck, resilient organizations anticipate, analyze scenarios, listen to those on the front line, test their defenses and learn from what almost went wrong. It is this approach that strengthens Safety Culture and makes it possible to learn before the damage happens.

And to do that, you don't just have to cross your fingers, you must cross your data. Data tells stories, reveals patterns, exposes weaknesses and indicates where to act. But this is only possible when there is the capacity to read, interpret and, above all, respond. A response that is not rushed, but agile, thoughtful, connected to the context and culture of the organization. It is this response that builds resilience. Because an organization that knows how to interpret its data and turns learning into action not only reduces work accidents, but it is also better prepared to deal with them when (and if) they happen.

Building a mature and resilient Safety Culture means, above all, courageously accepting that failures can happen, and therefore intentionally preparing our systems (people, processes, technology, culture) to deal with them safely. This is not a sign of weakness, but a gesture of lucidity and active care. It means:

- Continuously invest in strengthening the four capacities that underpin resilience.

- Create layered defenses, recognizing that no single barrier is sufficient on its own.

- Broadening the view of safety management means moving away from an exclusive focus on failures and punishments to a deeper understanding of how success is built daily. It implies understanding and strengthening the conditions that make safe work possible, such as adequate resources, sufficient time, continuous training, genuine spaces for dialog and active and attentive listening, especially in environments marked by constant pressure and variability.

This change in perspective, from trying to eliminate all errors to building the capacity to deal with it, profoundly transforms the way we conduct leadership, engage people, design processes, investigations and even the metrics we use to measure safety.

When we come to see safety as the presence of living capabilities and cultivate the goal of "failing safely", we make room for a more humane, proactive, realistic and ultimately much more effective approach to protecting who matters and what matters in the long run.

FOR REFLECTION:

1. **Recent near misses:** Think back to a near miss or minor incident that recently occurred in your area. Was it an example of "being lucky" or "having the structure to deal with the unexpected"? Which barriers really worked? Which ones failed? And, above all, what was learned, or left out?

2. **Dependence on luck:** Are there any processes or areas in your organization where safety seems to depend more on luck or individual effort than on a well-structured system? What could be done to turn this into something more predictable, more secure and less vulnerable?

3. **Learning from success:** Does your organization also observe and learn from successes? Especially in those moments when something unexpected happened but was handled well? How are good practices, well-made adjustments and the right decisions recognized and shared?

4. **Investment in capabilities:** How does your organization strengthen the four essential capabilities in practice: Anticipate, Monitor, Respond and Learn? Where is this already visible? Where is there still room to grow?

6

THE HEART OF CULTURE

PEOPLE, COMPETENCE AND ACTIVE CARE

Safety Culture, in its deepest essence, revolves around people. It is they, individually and as a team, who experience real work, identify risks, make decisions in real time, operate systems and, above all, choose to take care of each other daily.

No technology, no matter how advanced, no procedure, no matter how detailed, and no management system, no matter how structured, can replace something that is exclusively human. It is about people's unique ability to adapt continuously, judge sensitively, act creatively and sustain resilience in the face of the unexpected.

Strengthening the People pillar cannot be seen as just another item on the agenda. It is, in fact, the fundamental basis, an essential condition for any safety culture to reach maturity, be effective and stay alive over time.

CHANGING THE LENS: PEOPLE AS A SOLUTION AND SOURCE OF RESILIENCE, NOT JUST A SOURCE OF ERROR

Historically, many approaches to safety, perhaps influenced by more rigid engineering or management models, ended up treating the human factor predominantly as a source of failure. The focus was on controlling unpredictable, correcting behavior and reducing errors. The worker was considered the weak link in the system, someone who had to mold themselves to the prescribed procedures at the risk of compromising the whole.

Although this way of thinking has made some progress, it has proved to be quite limited. More recent studies in human factors, organizational psychology and resilience strongly suggest that we urgently need to abandon this view as the dominant one.

The current perspective invites us to invert this logic, recognizing people no longer as a problem, but as a fundamental element of solutions. They are essential sources of adaptation, practical intelligence, rapid responses and crucial decisions in the face of everyday complexity.

After all, the same human characteristics that can lead to error under certain conditions, such as intense focus, simplification in the face of complexity and adaptation under pressure, are also those that prevent countless work accidents daily. It is precisely these human impulses that discover safe paths, anticipate failures, balance risks and protect the system with sensitivity and experience.

People are not the weak link but are almost always the link that sustains the system. And it is precisely with this in mind that we can build a culture in which care is valued, listening is active, and safety is no longer just control, but trust.

Heroism and initiative: The ability to act quickly and decisively in the face of imminent danger.

Compassion and care: The genuine impulse to protect others, to intervene when something doesn't seem right.

Imagination and creativity: The ability to find solutions when procedures aren't enough.

Adaptation and flexibility: The ability to adjust work to actual conditions, even when resources are scarce.

Learning and improvement: The natural willingness to learn from experiences and transform everyday life based on them.

SEEING PEOPLE AS A SOLUTION THEREFORE MEANS:

Valuing practical experience: Recognizing that those on the front line carry living and irreplaceable knowledge. These are the people who know the real risks, the daily challenges and the solutions that really work in practice.

Active listening: Creating a culture where suggestions, warnings and concerns are listened to carefully and taken seriously. Where there is space to speak and, more than that, to be considered. Listening that generates a response, not just a record.

Involve in co-construction: Include those who carry out the work in the design of solutions. Bring the front line into the conversation, to participate in the creation and improvement of processes, tools and routines, not as guests, but as protagonists.

Contextualizing the mistake: Understanding that, in most cases, mistakes are not the result of individual negligence or incompetence. They reflect deeper issues, such as processes that don't work properly, tools that hinder rather than help, insufficient training, targets that put excessive pressure, communications that don't go where they should, or even an organizational culture that punishes those who try to get it right.

Investigating error from this perspective means asking the right question, trying to understand why a particular decision seemed reasonable to that person, at that moment and in that context. It is about analyzing the whole scenario, not just the person who was at the center of the situation.

This change in perspective does not cancel out individual responsibility, nor does it invalidate the importance of clear rules. On the contrary, it brings justice, depth and, above all, greater efficiency. Because managing safety with maturity is not about trying to control human beings, but about recognizing their value, their practical intelligence and their essential role in creating safer and more humane environments.

BUILDING COMPETENCE: BEYOND TRAINING, TOWARDS CONTINUOUS AND CONTEXTUALIZED DEVELOPMENT

Competence is undoubtedly the foundation on which safety is built, both individually and collectively. But it must be recognized that being competent in safety goes far beyond having taken part in initial training or undergoing periodic refresher training aimed solely at meeting legal or corporate requirements. These training courses are important, yes, but on their own they are far from preparing someone for the real challenges of everyday life.

True competence is not an end point. It is a living, continuous process, built into practice and adjusted all the time as work changes. It is not enough to know the "what" (the rules) and the "how" (the procedures). you need to cultivate the "why", the purpose behind each action, awareness of the risk involved, clarity about the impact of a choice.

More than just memorizing answers, it is essential to develop critical and reflective thinking. It means preparing people to make wise decisions in the face of unforeseen situations. It means strengthening their responsible autonomy and intelligent ability to adapt. After all, this is precisely what underpins safety in everyday practice, in the living and dynamic reality of the workplace.

HOW CAN WE BUILD THIS ROBUST AND CONTEXTUALIZED COMPETENCE?

1. **Onboarding as a fundamental and universal ritual:** The on**boarding** process is not just a formal step. It is one of the most important rituals in building Safety Culture. Whether you're joining the organization, taking on a new

role, changing areas or returning after a period away, onboarding is the gateway to the way we take care of each other here.

2. **That is when we have a real chance:**

- Clearly communicating the values that underpin our Safety Culture.
- Align expectations about behavior, responsibilities and commitments.
- Present, in a realistic and close-to-practical way, the risks involved in that role.
- Show, with attitude, that the leadership is present and committed to well-being.
- Reinforce the importance of open communication, reporting, listening and active care.

When integration is rushed, limited to bureaucracy or restricted to only certain levels of the organization, this also sends out a message. And what that message communicates can be dangerous, suggesting that safety is just a formality.

On the other hand, integration carried out with care, constant presence, active mentoring, close accompaniment and open dialog represents a valuable investment. It is precisely at this initial moment that the sense of belonging begins to build. And where there is true belonging, care naturally flourishes.

3. **Training beyond compliance: Engagement and critical skills:** Formal trainings need to stop being just about delivering content to meet requirements. To generate real transformation, they need to touch, involve and make sense in practice. It is time to evolve from the passive format to experiences that connect with the daily lives of those who live the real work. Effective training is training that:

Engaging: Brings interaction, promotes real exchanges. It uses case studies, simulations, dynamics and tools that invite participation and active involvement.

Practical and relevant: Speaks the language of those at the cutting edge. It uses everyday examples, connects with real challenges and shows that safety isn't theory, it is action.

Focused on "Why": Goes beyond the rule and shows the meaning. It helps us to understand the risks that are being controlled, bringing awareness to what lies behind each procedure.

And more than technical content, the focus also needs to be on developing critical skills, such as:

Risk Perception: Knowing how to identify danger, even in the most familiar routines.

Decision-making under pressure: Clearly choose the safest course of action when faced with difficult or unexpected situations.

Assertive communication: Having an active voice with respect, to warn, intervene, ask questions.

Teamwork and coordination: Collaborate conscientiously to ensure collective care.

Problem-solving: Analyzing scenarios and finding safe solutions, even in the face of the unexpected.

4. **Learning on the job and mentoring:** Much of what we learn about safety doesn't come from slides, but from experience. What we see, what we feel, what we share with those who have already walked the path. It is in practice, on the work floor, that competence takes shape. That is why it is essential to create an environment where this continuous learning is intentional, structured and supported.

Mentoring or sponsorship programs: Connecting newcomers or those taking on a new role with experienced and respected colleagues. Not just to teach how things are done, but to welcome, accompany, guide and inspire by example.

Regular safety dialogues: Real conversations, where the work happens. Spaces where leaders and teams talk about the risks of the day, the challenges of the task and the good practices that emerge in the process. Without excessive formality, but with listening, presence and intention.

Success-focused task analysis: Observing work not just to find faults, but to understand what works. Valuing successful adaptations, learning from them and spreading this living knowledge throughout the team.

Constructive and continuous feedback: Creating a culture where it is natural and safe to talk about what needs improving, but also to recognize what has been done well. Where feedback doesn't point the finger of blame but paves the way for growth.

2. **Developing non-technical skills (soft skills) as a pillar:** Safety is not only sustained by technical mastery or compliance with procedures. It truly flourishes when relationships are healthy, when there is listening, when care is also built into the way people communicate, connect and support each other.

3. The so-called "soft skills" are, in practice, essential safety skills. Developing assertive communication, teamwork, collaborative conflict resolution, sensitive and adaptive leadership at all levels and, above all, the ability to create an environment of psychological safety is what allows people to feel safe to talk, ask questions, intervene and learn together.

4. Building competence is a never-ending process. It is dynamic, lively and multifaceted.

5. The organization needs to go beyond compulsory training, hours accounted for or certificates handed out. It is necessary to invest purposefully and intentionally in the continuous development of people, both technically and humanly, deeply connecting learning to the day-to-day reality of work and to the greater purpose of caring for each other in a genuine and sustainable way.

GENUINE ENGAGEMENT: BEYOND CONFORMITY, TOWARDS COMMITMENT OF HEART AND MIND

Competence alone does not guarantee a lively, vibrant and proactive safety culture. We can have people who are technically prepared, who know the procedures, how to carry out the tasks and master the operational details. But if they aren't truly engaged, if they don't see safety as a value that makes sense, that touches them, that belongs to them, safe behavior will be fragile, irregular and dependent on the constant presence of someone demanding it.

When safety is seen only as an obligation, care becomes mechanical. What is asked is done, but without presence, without conviction, without depth. And that makes everything vulnerable, because there is no consistency when the motivation is external.

Real engagement doesn't come by decree or threat. It needs to be cultivated with intentionality. It needs to be awakened by connection, connection with purpose, with the other, with the value of going home well and making the other go home too. This is how hearts and minds align with the cause of safety. And that is when it really becomes strong.

HOW DO YOU NURTURE THIS DEEP AND LASTING COMMITMENT?

1. **Shared and connected purpose:** True engagement is born when people understand why they are doing what they are doing. When safety is presented only as a "rule", it may generate compliance, but it hardly generates connection. You must go beyond that. You must touch on what really matters.

Personal values: Showing that safe practices exist to protect everyone's life. Safety is about getting home safely, hugging those you love, living life with freedom and health. Real stories, told with respect, have the power to generate identification and raise awareness.

Caring for your colleague: Reinforce that safety is a collective commitment. That every safe attitude is also a gesture of care for those around you. It is not just about me; it is about all of us. It is known that when one person takes care, the whole team grows stronger.

Organizational objectives: To show how safety supports business continuity, protects the company's reputation, strengthens the quality of what we deliver and, above all, reinforces the pride of being part of something that really cares for people.

Leadership plays an essential role in this process. Not just by communicating, but by living with this purpose consistently, in every choice, in every conversation, in every decision. Because when safety stops being just a priority and becomes a non-negotiable value, it takes root. And it becomes part of who we are, not just what we do.

2. **Active and meaningful participation:** True engagement is born when people realize that they have a voice, and that this voice is heard, considered and taken seriously. Safety cannot be imposed on workers. It needs to be built with them, through dialog, listening and practice.

Involvement in risk identification: Create accessible, simple and reliable channels for the front line to report hazards, unsafe conditions and ideas for improvement. And, more than that, ensure that these contributions receive feedback and generate action.

Participation in investigations: Include professionals who live the real work in investigations into incidents and near misses. Not as culprits, but as experts in what they do every day. They are the ones who know the details, the deviations, the pressures and the adaptations. And it is with them that solutions become possible.

Co-developing solutions: Invite workers to review procedures, adjust tools and co-create programs. In this way, solutions don't just look good on paper, they make sense in practice.

Committees and working groups: Creating participatory spaces where different areas and levels can discuss relevant issues, propose improvements and, with autonomy, put action into practice. When there is trust and space to build together, there is transformation.

Participation generates belonging. And where there is belonging, true commitment is born. Because whoever helps to build also helps to look after. And they defend with their hearts what they have helped to create with their own hands.

3. **Meaningful and fair recognition:** Recognizing what is done with awareness, care and intention is a powerful way of strengthening the culture we want to see grow. When we celebrate safe behaviors, good practices, suggestions put into action and, above all, gestures of active care, such as interventions between colleagues, we are reinforcing the right path with lightness and truth.

Specific: Aimed at specific behaviors. It is not about generic praise, but about clearly naming what has been done and why it matters.

Timely: Happening as close as possible to the moment when the behavior was observed. Delayed recognition loses strength and connection.

Genuine: It needs to be true, heartfelt, without obligation, without protocol. People notice when it comes from the heart, and that makes all the difference.

Varied: It can come in many forms: a public thank you, a mention at a meeting, a handwritten letter, a symbolic tribute, a team celebration. Recognition doesn't have to be expensive; it has to be human. And care must be taken not to fuel competition or focus only on cold indicators, such as "zero work accidents".

The focus of recognition should be on effort, attitude and daily commitment, not just the absence of negative events. Because what sustains safety, in practice, are constant small gestures. And when they are seen and valued, they multiply.

4. **Psychological safety: The indispensable foundation:** Perhaps the most critical pillar, and at the same time one of the most delicate to build for true engagement, is psychological safety. This is the feeling that it is safe to be who you are in the workplace. That you can speak up, ask questions, admit, disagree and contribute without fear of being ridiculed, punished or silenced. it is this environment that allows people to:

- Ask questions, even those that seem too simple.
- Admit mistakes or difficulties without fear of judgment.
- Share concerns and disagreements with respect and openness.
- Have the courage to propose new ideas or challenge what is already established.

Without this safe space, fear takes over. Failures go unreported. Risks are hidden. Silence sets in. And learning stops. Engagement becomes a facade, and Safety Culture loses depth.

Guaranteeing psychological safety is not something that can be solved with a one-off action. It is an ongoing construction that begins and is sustained by leadership, it requires humility to listen, the openness to recognize limits, the willingness to learn from failures and the courage to show vulnerability naturally.

Psychologically safe environments are not born by chance. They are cultivated by leaders who choose, every day, to respond with empathy, welcome with respect and care with truth. Because only where there is trust does care truly flourish.

ACTIVE CARE: THE ULTIMATE EXPRESSION OF INTERDEPENDENCE AND VALUE FOR LIFE

In Safety Cultures that reach higher levels of maturity, such as the interdependent stage described by the Bradley Curve or other similar models, care goes beyond both individual responsibility and simple compliance with rules. It becomes something much deeper and more consistent, turning into a genuinely shared commitment, where collective well-being is experienced as a real priority, going far beyond words or speeches.

In this context, people don't just care about themselves. They develop a real sense of responsibility towards others. There is connection, belonging and genuine affection. This culture finds its most authentic expression in active care, manifested whenever someone respectfully and courageously chooses to intervene in critical situations.

Intervene when you see a colleague at risk, when you identify something that could go wrong, or when you feel that someone is not well enough, either physically or emotionally, to perform a task safely. It is a gesture that arises naturally from the strength of the bonds created and not from a mere formal obligation. It arises from the clear awareness that safety, in practice, is always a collective construction, maintained daily with attentive eyes, present attitudes and hands ready to help.

IT IS CRUCIAL TO DIFFERENTIATE ACTIVE CARE FROM OTHER DYNAMICS:

- **It is not "watching" or "policing":** Active care is not about monitoring or pointing out mistakes. The focus is on protecting, on looking after someone in the team because it is about life, not perfection.

- **It is not about "snitching" or "gossiping":** The intention is not to expose anyone, but to prevent anything that could cause harm. Whenever possible and safe, the conversation should be direct, respectful and welcoming. When necessary, it can be taken to the leader or safety department, with the focus on the solution, not the accusation.

- **It is not an act of superiority:** Intervening isn't about showing you know best. It is about caring. It is about saying, with attitude and respect: "i care about you". Because in a culture of true maturity, caring for others is not an invasion, it is a commitment.

Active care is the living expression of interdependence. It is when, in practice, we assume that the other person's safety is also my responsibility, just as mine is theirs. It is a gesture of courage and maturity, because it requires going through the discomfort of intervening, of appearing invasive, of generating possible discomfort. But above all, it is an act of affection and respect.

Intervening requires sensitivity: knowing how to speak with empathy, focusing on behavior and risk, and never on the person. It means arriving with presence, not judgment. It means saying "I care", even if only in a few words.

On the other hand, it also requires humility. Receiving a warning from a colleague without closing yourself off, without feeling attacked. Recognizing that there has cared there. That the other person saw something that I may not have seen and wanted to protect.

In a culture that has reached maturity, this form of exchange is welcomed exactly as it should be, that is, as a valuable gift. A constant reminder that we are all in this together, that no one needs to or should walk alone when it comes to safety.

CULTIVATING AN ACTIVE CARE ENVIRONMENT:

Active care doesn't happen by chance. It needs to be cultivated, with intention and constancy, throughout the organization, especially by the leadership. It is a process that requires building a favorable ecosystem, supported by pillars such as:

1. **Mutual trust and psychological safety:** As we have seen throughout this chapter, no one feels comfortable intervening or receiving open feedback if there is no real safety between people and with the leadership. Psychological safety needs to be experienced daily; it is not enough to be written down in values. Where there is fear of embarrassment, judgment or retaliation, care is silent. And when care is silent, the risk grows.

2. **Communication and feedback skills:** We can't expect people to know by instinct how to intervene sensitively and clearly. The organization needs to teach. Develop assertive communication skills, active listening and, especially, safe and respectful practices for giving and receiving feedback. Simple tools, such as: "I have observed [behavior or condition] ... I'm concerned about [risk]... Can we talk about [suggestion or solution]?" can help turn a difficult intervention into a caring gesture that brings you closer together.

3. **Normalizing and valuing intervention:** The role of leadership is to make care visible and legitimate. Intervening, when necessary, yes, but above all thanking and publicly recognizing those who have had the courage to care. When intervention is seen not as criticism, but as an expression of mutual responsibility, it ceases to be an exception and becomes part of the natural way of doing things.

4. **Clarity about the process:** It is essential that everyone knows how to intervene, when to intervene and where to go for support if direct intervention is not possible. Having clear, secure and fair channels for reporting a concern strengthens trust in the system. A transparent process conveys respect, and where there is respect, care gains room to grow.

Advice for leaders: When you witness an act of active care, acknowledge it immediately and, if appropriate, make it public. Something simple but meaningful can be said as follows:

"Thank you, [name], for your courage and attention in talking to [colleague] about [situation]. It is precisely this attitude of mutual care that strengthens our safety as a team."

This sincere recognition reinforces the organization's values, inspires others and demonstrates in practice that caring is a genuinely valued behavior.

Strengthening the people pillar is an ongoing commitment. It is an investment that transcends the technical aspect, as it involves developing human skills, cultivating engagement based on a greater purpose, creating real spaces for participation and making a conscious and constant effort to sustain an environment where mutual care can flourish.

Deeply recognizing that safety's most asset is not systems, technologies or procedures, but the people themselves, it is essential. They are capable of perceiving details that sensors don't pick up, of adapting where scripts fail, of learning from their experiences and, above all, they are the ones who choose to take care.

In the next chapter, we will take a deeper look at the Processes and understand how they can and should be designed not to limit, but to support, protect and enhance human capacities. Because when the system values and strengthens care, the heart of Safety Culture pulses with much more strength, safety and resilience.

FOR REFLECTION:

1. **Climate of trust:** On a scale of 1 to 5, what is the level of trust and psychological safety in your team so that people feel truly comfortable intervening with each other when they perceive a risk?
2. **Feedback skills:** Does your organization offer specific training that teaches, in practice, how to give and receive constructive feedback on safety in a clear, respectful and welcoming way?
3. **Leadership example:** How **do** the leaders in your area demonstrate on a daily basis that active care is important? How do they encourage this behavior in their teams?
4. **Barriers to intervention:** What are the main barriers hindering the practice of active care in your organization today? Are they cultural, procedural or individual? What still needs to be changed so that intervening becomes something natural and valued?

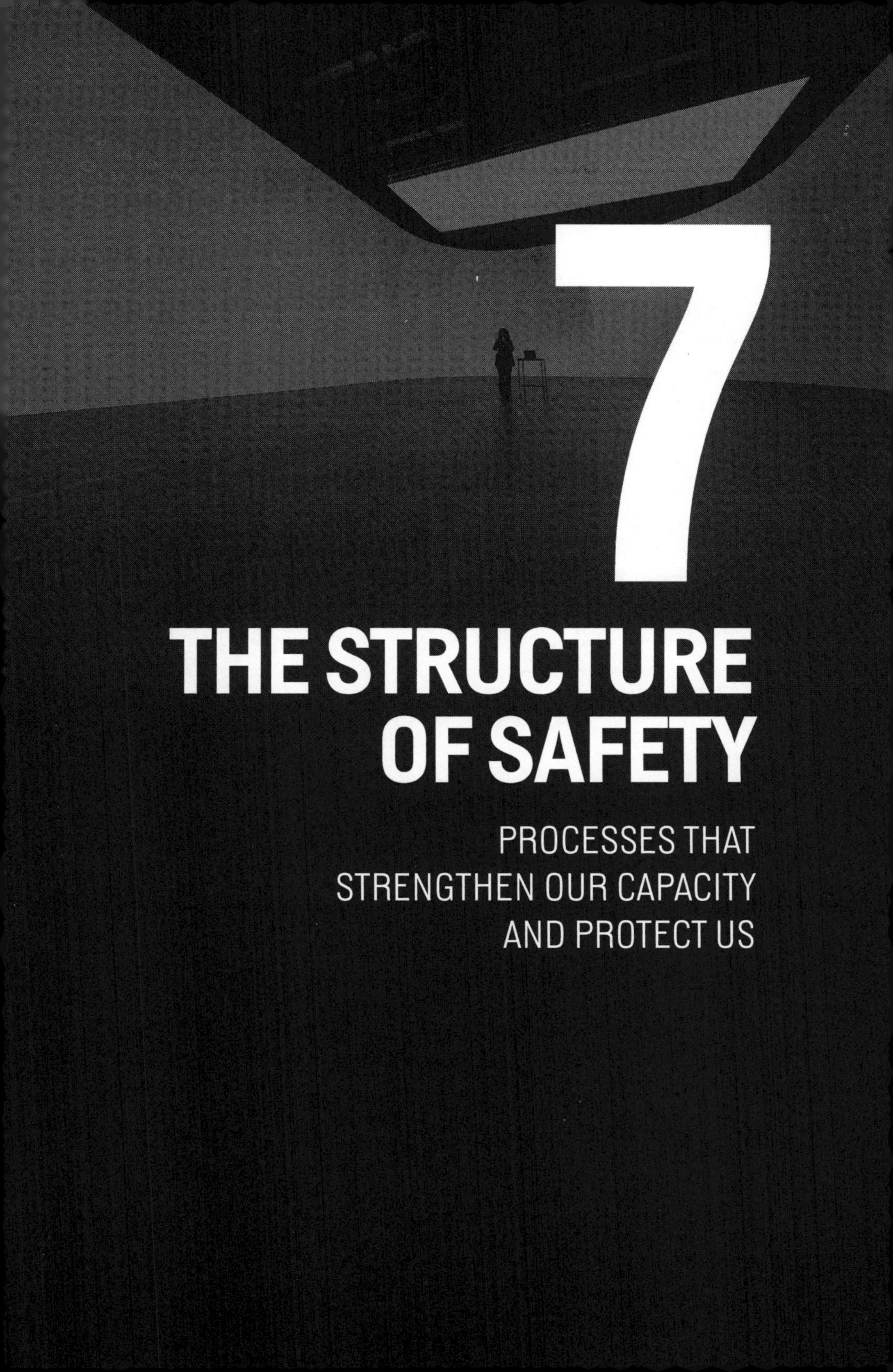

7

THE STRUCTURE OF SAFETY

PROCESSES THAT STRENGTHEN OUR CAPACITY AND PROTECT US

If people represent the beating heart of Safety Culture, with their unique capacities to adapt, learn and care, processes correspond to the skeleton and nervous system that provide support, direction and connection to this culture.

When processes are carefully designed, aligned with the day-to-day reality of the workplace and aimed at strengthening human capacities, they create a solid foundation, organize the journey and concretely materialize what we believe about safety. Through them, values are manifested daily, translating into consistent actions that support care and effectively prevent risks.

However, the mere existence of processes does not guarantee protection. When they are drawn up in a distant, overly complex way or imposed just to fulfill an obligation, they can have the opposite effect. Instead of supporting, they limit. Instead of facilitating, they become obstacles.

Processes that don't dialogue with practice generate frustration, encourage dangerous improvisation, undermine trust between the front line and management and stifle what we most need to preserve: the capacity for initiative, adaptation and learning.

It is in this context that phenomena arise such as paralysis caused by over-protection, in which excessive bureaucracy blocks action, or paralysis caused by rigidity, which prevents adjustments and inhibits continuous improvement.

When a process deviates from its original purpose and no longer serves real work, it loses its essence. So instead of protecting, it ends up weakening the system. That is why it is more important than simply creating rules to build paths that support the caregiver. We need processes that serve life, far beyond paper.

FUNDAMENTAL PRINCIPLES FOR SAFE, RESILIENT AND HUMAN-CENTERED PROCESSES

For processes to stop being just formal requirements and become true allies of safety and resilience, it is essential that their design, implementation and ongoing care are guided by principles that put people, and the real work they do, at the center of everything. These principles, inspired by the lessons of highly reliable organizations and lessons learned about human factors, remind us that safety doesn't live on paper, but in lived practice.

1. **Unwavering focus on real work:** This is undoubtedly one of the most important principles and, paradoxically, one of the most often forgotten. Effective processes are not generated behind a desk, nor are they based on idealized assumptions about how work should take place. To be truly efficient, they need to reflect what happens on the ground where everything takes place in the field, on the Gemba, on the shop floor, on the road, at the service front.

2. Real work brings with it variability, pressures, interruptions, limitations and adaptations. It is different from what is in the manual and often also from what is imagined by those far from the operation. And it is precisely in this difference that the risk lies, but also the potential for learning.

How to do it: You must go to where the work is taking place. Observe with respect, ask with curiosity, listen with humility. Operators are the true experts on reality, they are the ones who know where the process is tight, where the tool fails, where you have to improvise to get it right.

Task analysis needs to go beyond a simple step-by-step description, also capturing the successful adaptations, discrete deviations and practical adjustments that make work possible and safe. In addition, it is essential that feedback from the front line is listened to seriously and incorporated directly into process reviews. Ignoring this gap between what is written on paper and what actually happens daily means compromising safety as a whole sooner or later.

Example: A maintenance procedure presupposes ideal conditions. But on a day-to-day basis, technicians encounter corroded parts, limited access, tools that don't work as they should. When the process focuses on what happens, it stops forcing silence about deviation and starts offering practical guidance, safe alternatives and clear decision points.

Because processes that recognize reality don't punish adaptation, they welcome it, organize it and strengthen it. And that changes everything.

3. **Simplicity, clarity and radical usability:** Unnecessary complexity is a silent enemy of safety. Procedures that are too long, full of technical jargon, poorly structured, visibly confusing or difficult to access at the right time don't help, they hinder. Under pressure, this type of document gets lost. And the risk increases.

4. Usability is not a detail. it is a commitment to those who are there, carrying out the task. It is a care that protects. A procedure is only useful if it is understood, applied and, above all, if it makes sense to those who need it in practice.

How to do it:

> **Write clearly.** Direct sentences, accessible language, active voice.

> **Highlight what is essential:** critical steps, main risks, points of attention.

> **Avoid excess.** Details that don't help, confuse.

> **Use visual and objective formats:** clear checklists, intuitive flowcharts, tutorials with images or short videos.

> **Position information where it is needed:** at the point of use, at the right time.

And, above all, test it with those who will be using it. Before finalizing it, give it to someone who is doing the task, but who isn't involved in creating it, and ask them to interpret the steps. You'll quickly find out where the doubts lie.

Practical advice: The aim is not to create a beautiful or technically flawless document. The aim is to genuinely support the safe and effective execution of work. If a procedure hinders more than it helps, it ceases to be protection and becomes a barrier. Safety does not go hand in hand with bureaucracy. It goes hand in hand with clarity, lightness and practicality at the service of life.

5. **Deliberately building capabilities and defenses in depth:** As we saw in Chapter 5, one of the pillars of a resilient culture is recognizing, with humility and realism, that failures happen. Even with the best training, the best intentions and the best-designed processes, human error and technical failure will still be part of the picture. And not because of negligence, but because we work in complex systems, with multiple variables and daily pressures.

6. Therefore, processes should not only be designed to prevent the initial error. This is important, of course, but the greater focus needs to be on something even more effective and powerful. We must create layers of protection capable of detecting, absorbing, containing and enabling recovery before a failure becomes real damage.

7. This concept, known as barriers in depth, reminds us that relying on a single barrier is a risk. We need multiple, diverse, complementary and intentional layers. That way, if one fails, others will be there to protect.

How to do it:

- Create redundancies, backup systems, secure alternatives.
- Implement cross-checks such as double-checks at critical stages.
- Use early warning systems, reliable alarms, visual indicators, early signals.
- Include physical containment mechanisms (such as containment basins) or logical ones (such as interlocks).
- Ensure well-defined contingency plans that are known and practiced by everyone, and not just documents kept in drawers.
- Diversify defenses by combining human, technical and procedural elements so that a single failure does not compromise the system.

Designing with this in mind means accepting that mistakes can happen and, even so, making sure they don't cause pain. It means protecting with structure. It is creating systems that welcome failure, learn from it and don't collapse in the face of the unexpected.

Failing safely is not a sign of weakness. it is a sign of maturity. It is the mark of a culture that understands that perfection doesn't exist, but consistent care does.

8. **Intelligent application of the hierarchy of controls:** When defining control measures within processes, rather than simply listing solutions, it is essential to make conscious choices and prioritize what offers consistent protection. The hierarchy of control guides us along this path, helping us to see that not all controls offer the same level of safety.

It invites us to start with what is strongest, most reliable, least prone to human error, and to follow responsibly when it is not possible to eliminate the risk.

Elimination: Remove the danger at source. This is the ideal solution, when possible, it is the safest way.

Substitution: Exchanging the process, material or practice for something less dangerous.

Engineering controls: Create physical barriers between the person and the risk. Modify the environment, equipment, systems.

Administrative controls: Changing the way people work: Rules, procedures, training, work organization.

Personal Protective Equipment (PPE): The last barrier. Individual, dependent on correct and constant use.

When we rely too much on the weakest controls, those that depend on human behavior always working perfectly, we increase the system's vulnerability without realizing it.

This is why processes need to actively and responsibly seek out controls that offer the most protection and the least margin for error. Whenever it is technically and economically feasible, the choice should be to move up the hierarchy and not settle at the lowest levels.

Protecting with awareness is exactly that. It means looking at reality as it is, respecting its natural limits, but never failing to look for the options that are most caring and empowering. Safety is built through choices that support and empower people, starting with the way we design each stage of the process.

9. **Learning and continuous improvement as DNA (living PDCA cycle):** Processes are not meant to gather dust on shelves or get lost on servers. They need to be alive, updated, observed, adjusted and, above all, connected to practice. A process that doesn't learn from experience ceases to be a support and becomes an obstacle.

10. That is why the Plan-Do-Check-Act (PDCA) cycle needs to be more than just a pretty picture on the wall. It needs to happen, with intentionality and depth, as part of the routine that underpins safety.

Plan: Develop or revise the process based on everything we've seen so far, with special attention to the work as it happens.

Do: Implement with clarity, ensuring that people understand and don't just receive the content. Simple communication and training connected to reality make all the difference.

Checking: True monitoring means carefully observing what is working, what is being ignored and the small, silent deviations that occur daily. It involves observing carefully, listening attentively, gathering genuine feedback from the front line, analyzing indicators in depth and, above all, continuously learning from everything that happens with incidents, near misses and also from what works daily, despite the difficulties.

Act: Use what has been learned to adjust. Correct what isn't working, simplify what is complicated, update what has become detached from practice. And then start the cycle again with lightness, presence and truth.

Effective processes are not those that follow a closed pattern. They are the ones that grow with the organization. They are those that adapt to what reality teaches them, without losing their commitment to care. Continuous improvement is a sign of maturity. It is knowing that there's always something to learn, and that caring is also about having the courage to change.

GENUINE ORGANIZATIONAL LEARNING MUST SYSTEMATICALLY FEED THE EVOLUTION OF PROCESSES.

Processes that make-or-break Safety Culture.

Not every process is neutral when it comes to safety. Although all operations bring some level of influence, there are management paths that have the power to shape or break what we believe, practice and sustain as an organizational culture. These processes speak. And they often speak louder than any speech, goal or written policy.

It is on a day-to-day basis that they reveal what really matters to the organization. And it is from the way they are conducted that workers understand whether safety is just a checklist item or a genuine value. Let's take a closer look at one of these processes together.

1. Risk analysis and management (Beyond Compliance):

What it is: It is the formal and intentional process of identifying hazards, evaluating risks in terms of their severity and likelihood, and deciding, with responsibility, which control measures will be put in place.

Cultural impact: When this process is done well, it clearly communicates that the organization takes the lives of its employees seriously. That it is not content with just complying with the minimum required by law, but wants to go further, with awareness and responsibility.

It means looking closely at risks that could cause serious or fatal injuries, even if they are rare. It means understanding that low frequency does not mean low importance. It means applying tools that really help you think, such as APR, HAZOP, Bow-Tie, HRN and, above all, it means giving a voice to those on the front line. Because these are the people who live with the risks daily and know the details of the operation like no one else.

More than drawing controls on paper, it is about ensuring that they work in practice and remain effective over time.

Warning sign: When this process becomes a formality, it loses its power. When it is done only to "fulfill a schedule" or ignores those who really live the operational routine, it does not protect. It deceives.

Shallow analyses, weak or absent controls, or decisions that are disconnected from the reality of those who carry them out are symptoms of a fragile, reactive and complacent culture. And, unfortunately, they are also fertile ground for serious work accidents.

Safety that transforms is not based on words. It is built in practice, and it all starts with how we choose to look at risk.

2. Change management: The guardian against hidden risks:

What it is: It is the formal process for assessing and controlling the risks involved in any type of change, whether in equipment, technology, procedures, raw materials, structure or people.

Cultural impact: When applied well, the change management process shows that the organization recognizes that all change carries risks. Even well-intentioned changes can weaken existing controls or create new vulnerabilities.

A consistent process ensures that any change is carefully evaluated before it is put into practice. This includes defining controls, reviewing procedures, training teams and communicating clearly to everyone involved.

Warning sign: Changes made without prior analysis or with controls applied after the process has already begun are signs of carelessness. Major industrial work accidents, such as Piper Alpha and Texas City, had change management failures as a contributing factor.

When this process is ignored or dealt with hastily, the message is clear: convenience has won out over safety.

Permission to work: The ritual that protects lives.

What it is: It is the formal system used to authorize and control high-risk activities, which are not part of the routine and involve significant hazards, such as hot work, confined space, heights, excavations and energy isolations.

Cultural impact: An effective Work Permit is not just a piece of paper. It is a ritual of communication, verification and authorization. It ensures that the risks have been assessed on site, that the necessary controls are in place, that everyone knows what can happen and how to protect themselves, and that someone responsible takes the decision to authorize the execution.

When this process is done well, it shows that the organization values caring for life and does not give up rigor, even in one-off tasks.

Warning sign: When the process turns into a bureaucratic routine, with permissions signed ahead of time, checks ignored or a rush to release the activity, an important warning **sign** goes off. This shows that the culture has become complacent, that risks are being minimized and that safety is no longer a priority.

3. Incident investigation and organizational learning (focus on solution, not blame):

What it is: It is the process that analyzes everything from small deviations to serious work accidents, with the aim of understanding their causes and preventing them from happening again.

Cultural impact: The way an organization reacts to incidents reveals a lot about its values. When the investigation only seeks to pin the blame, apply generic retraining or ask for "more attention next time", it creates fear, blocks report and prevents real learning.

Really effective investigations go much further. They take a fair and comprehensive stance, analyzing the system in its entirety, considering decisions made, pressures suffered, design flaws and possible gaps in procedures. They seek to deeply understand the reasons that led people to act in a certain way in that specific context. Based on this understanding, they build sustainable improvements capable of strengthening the defenses of the organization.

Practical advice Use methods that help you see the system, such as the 5 Whys for organizational causes, the Ishikawa Diagram or the analysis of causal factors (immediate, basic, root and contributing causes). And always focus on more effective controls, not just palliative actions.

4. Contractor management (extending the culture across borders):

What it is: It is the set of processes for selecting, integrating, monitoring and evaluating contractors working within the organization.

Cultural impact: The way an organization looks after its contractors reveals how much of a value Safety Culture really is. When there are rigorous criteria from selection, with safety as a priority, not just the lowest price, to onboarding, monitoring in the field and the final evaluation, it shows that everyone's life matters, regardless of the badge they carry.

Treating contractors with the same care and standards given to their own employees reinforces a mature culture that understands that responsibility cannot be outsourced.

Warning sign: When the organization simply asks for documents and expects the contractor to take care of safety on their own, or when they ignore unsafe practices so as not to interfere with the progress of the service, the message is clear: safety is secondary. And that weakens the whole culture.

5. Emergency preparedness and response (readiness for the unexpected):

What it is: Plans, procedures, resources and training to deal with emergency situations such as fires, leaks, serious work accidents and evacuations

Cultural impact: When these processes are well defined, communicated clearly, tested frequently through realistic simulations and reviewed responsibly, they show that the organization recognizes that failures can happen and is prepared to protect lives even in the face of the worst-case scenario.

This readiness communicates, without words, that safety is not a good-weather speech. it is a value that holds firm even in the most extreme situations.

Warning sign: Outdated plans, drills done only out of obligation, insufficient resources or undefined roles at the time of a crisis reveal a dangerous fragility. They show that, in practice, the organization is not ready, and that puts at risk everything it is trying to protect.

6. Safety audits and inspections (opportunities for dialog and learning):

What it is: Formal processes used to verify that safety standards and procedures are being followed and to identify practices or conditions that could pose risks.

Cultural impact: When well conducted, audits and inspections go beyond the checklist. They become moments for listening, observing real work and sharing experiences. They value what is working well, seek to understand the context of behaviors and make room for improvement based on facts, not judgments.

Prepared auditors, who know how to observe with respect and dialog with empathy, help to build trust and promote consistent change.

Warning sign: When audits are seen as punitive, aimed only at finding errors or applying blame, they lose value. If they don't generate learning or real actions for improvement, they weaken the culture instead of strengthening it.

These are examples of processes that, when taken seriously, transform. But when done out of obligation, they become obstacles to building a safe and lively culture.

THE INDISPENSABLE SYMBIOSIS: WEAVING THE NETWORK BETWEEN PEOPLE AND PROCESSES

At this point, it is essential to reinforce a simple and profound truth: people and processes are not separate or disputed paths within Safety Culture. They go hand in hand, influence and complement each other. The strength of culture lies precisely in this interweaving, as in a fabric in which no single thread holds together.

Processes empower people: When well-constructed, processes are pro-life tools. They guide, protect and organize real work. They are designed to support, not complicate. They provide safe boundaries, facilitate collaboration and reinforce the defenses that underpin day-to-day operations.

People bring processes to life: But processes don't gain momentum on their own. It is the attentive eye, critical judgment, experience and commitment of people that make them real. Without this, what has been designed is lost, weakened or becomes invisible. Processes without people are instructions without a soul.

How can we cultivate this living connection?

1. **Co-creation and genuine involvement:** The process needs to come from those who live it. Bringing in frontline workers to review, adjust and validate each step is the most effective way to ensure that everything is useful, applicable and respected in practice. They know where it hurts, where it fails and what can be done better.

Practical example: Are you going to revise a lockout/tagout (LOTO) procedure? Build the new flow with electricians, operators and mechanics who do the task every day. Listen to where the difficulties lie, validate the solutions with those who put their hands to the tool.

2. **Open, secure and responsive feedback** channels: Create simple, accessible and secure channels for people to report challenges and propose improvements. More important than receiving is responding. When a suggestion comes in, it needs to be taken on board, analyzed and, where possible, turned into action. Giving feedback gives value to the voice of those who live the work.

Practical advice: Set up a simple channel (physical or digital) for suggestions for improvement. And never leave a suggestion unanswered.

3. **Navigating tension: Adaptive flexibility vs. risk deviation:** Not every deviation is irresponsible. Sometimes it is the only way for workers to get the job done safely and with quality. But it is also true that some deviations are the result of haste, negligence or lack of care.

4. The maturity of the organization lies in knowing how to listen before judging, investigating openly and seeking to understand why the choice was made. Often, a deviation points not to the error of the worker, but to the failure of the process.

For reflection: Does your organization have a structured process for analyzing deviations with a focus on the context and the system? Or are you still stuck with the idea that all you have to do is find the culprit?

In the end, this is how a solid culture is built, through genuine listening, constant involvement and mutual respect between those who create and those who execute. When people and processes go hand in hand, in complete harmony, culture flourishes with deep and firm roots.

5. **Processes as support tools, not punitive controls:** The way workers see processes says a lot about the organization's culture. When they are perceived as tools that facilitate safe work, that make sense in practice and that have been well constructed, engagement grows. Adherence becomes natural and voluntary. But when processes seem distant from reality, bureaucratic, created by "those who have never set foot on the shop floor", or are used only to point out errors after an incident, they lose their value. The result? Resistance, deviation and disbelief become part of the routine.

Designing processes with a focus on simplicity, collaboration, real application and continuous improvement is what underpins safety. They should serve as a support, not a trap. And they only come to life when they are driven coherently by leadership, which is the pillar we'll look at next.

KEY POINTS OF THE CHAPTER:

- Processes underpin safety, but they need to be connected to real work and people.
- Essential principles: Focus on the work as it is done, simplicity and usability, defenses in depth, hierarchy of controls and continuous learning (living PDCA).

- Critical processes (risk analysis, change management, work permits, investigations, contractor management, emergencies and audits) strongly shape the culture.
- The connection between people and processes is essential and is built through co-creation, active listening, respect for adaptation and processes that support, not punishment.

PRACTICAL ADVICE:

Question works as imagined vs. work as it is done: Ask teams how work is actually done, not just how it is written.

Invest in usability: Simplifying can protect more than creating new rules.

Celebrate feedback: Making flaws visible is a contribution, not a criticism.

Audit critical processes: Personally observe how critical processes are being applied on a day-to-day basis.

FOR REFLECTION:

1. To what extent do your processes reflect the work as it happens in practice? Do they reflect the reality on the front line or are they still stuck to what is imagined on paper?
2. Are your processes perceived as allies in care or as bureaucratic obstacles that keep people away? Do they help to protect, or do they just require steps to be taken that make no clear sense?
3. How does your organization view the adaptations made by the teams?
4. Is there room to understand when flexibility comes from experience and when it hides real risk?
5. What ways exist today for the voice of those in the operation to be heard, welcomed and transformed into improvement? Does feedback from the front line find genuine listening or does it get lost along the way?

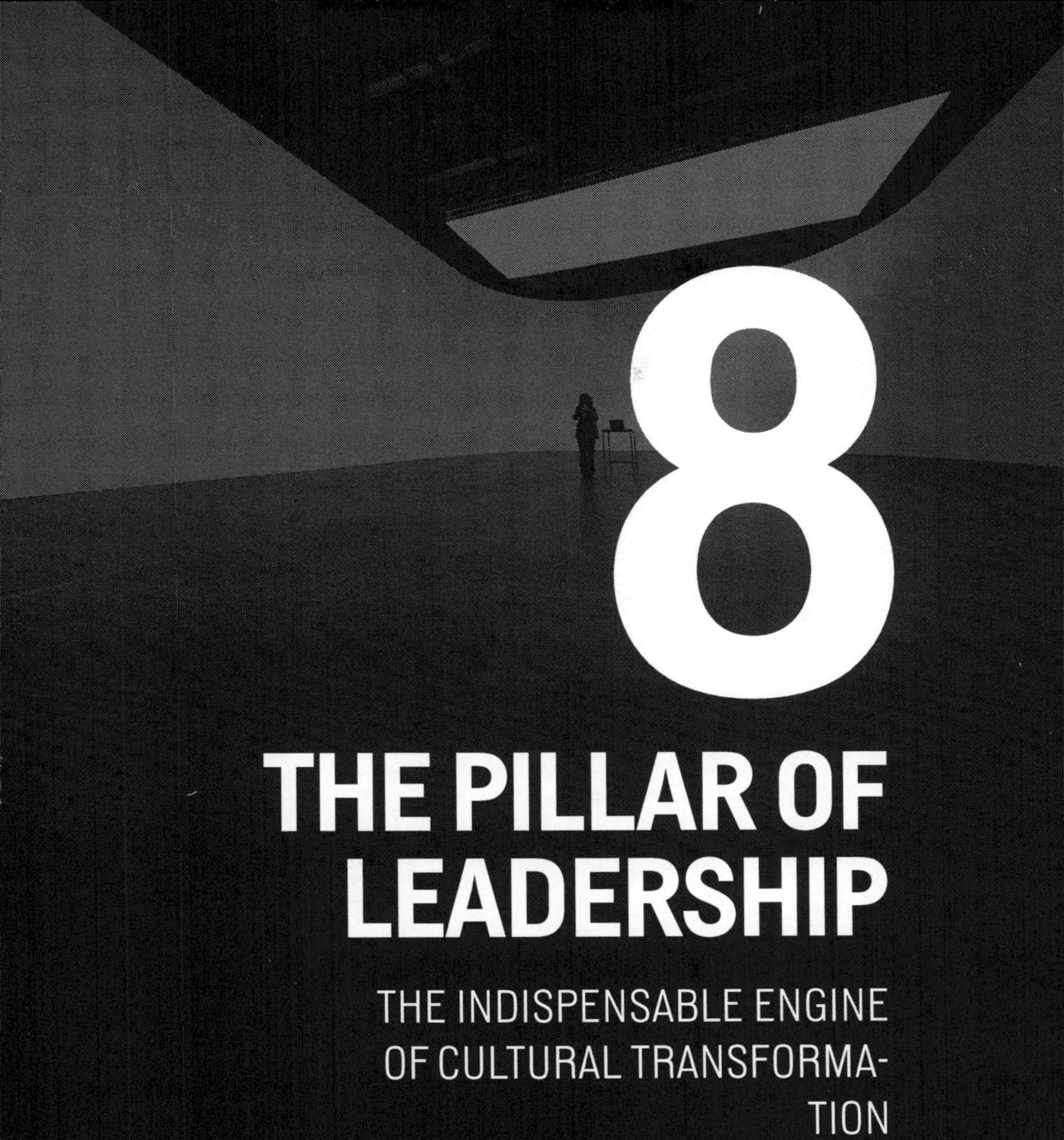

8

THE PILLAR OF LEADERSHIP

THE INDISPENSABLE ENGINE OF CULTURAL TRANSFORMATION

If people and processes are the pillars that support the structure of safety, leadership is the solid foundation that underpins all this construction. It is leadership that maintains, directs and strengthens the safety culture daily. In countless analyses of organizational successes and failures, one aspect is consistently revealed: leadership, present at all levels, is the essential element, the engine that transforms safety into a value that is genuinely lived, not just declared.

This responsibility cannot be seen as an additional function or something that can be delegated. It is part of the commitment of those who choose to care, those who understand that protecting lives is a daily priority.

Peter Drucker once wrote that leadership means raising people's vision to higher levels, while also pushing their performance to higher standards. In safety, this idea translates into inspiring an authentic commitment to care, clarifying precisely what is expected in each behavior, ensuring the availability of the necessary resources such as time, staff and the right tools and, above all, always acting consistently.

Safety is not a goal that changes according to circumstances. It is a deep value that reveals who we are and everything we choose to preserve.

THE DECISIVE AND MULTIPLIER IMPACT OF LEADERSHIP: THE CASCADE EFFECT ON CULTURE

The power of leadership in shaping culture is not empty talk, it is a reality proven by practice and solid data. Numerous studies show the profound impact that leaders' attitudes have on the choices and perceptions of their teams. This influence happens through subtle but powerful pathways that spread throughout the organization like a domino effect.

Mirror effect (behavioral modeling): The Leader as Moral Compass.

The fact: Most employees, more than 82%, carefully observe the behavior of their immediate leaders. This is one of the oldest and most natural ways of learning by example, according to the Deloitte Ethics Survey, 2023.

In practice: When a leader uses PPE correctly, follows procedures rigorously, including the so-called "Golden Rules", participates in safety talks with a real presence and shows care in their gaze and attitudes, they are teaching all the time. Their consistency becomes a compass for the team, which learns what is expected not just through words, but through experience.

The opposite also teaches, although it hurts deeply. When a leader fails to act in the face of a dangerous shortcut, ignores an obvious flaw or acts with disregard, the message transmitted is quite clear and damaging. With these behaviors, it becomes clear that safety is not as important as claimed. In this scenario, the discourse weakens, trust crumbles and risk finds room to manifest itself.

A real example: Paul O'Neill, former CEO of Alcoa, is a name that inspires me when it comes to leadership with purpose. When he took over the company, his first public speech was not about profit, expansion or innovation. He chose to speak firmly about safety. He set the goal of zero work accidents and showed, with consistent attitudes, that this was a non-negotiable priority. He visited the units, demanded reports within 24 hours and held managers directly responsible. This real commitment spread throughout the organization, transforming the culture and, as a natural consequence, also improving financial results.

Interpretation and validation of the rules: When the example confirms the value.

The fact: Rules only become meaningful when they are lived. It is not enough for them to be on paper. The way they are respected, or ignored, by leaders defines whether they will be followed or forgotten.

In practice: Employees observe attentively much more than we imagine. Around 80% of them understand what really matters from the way leaders act in practice. When managers follow the rules seriously, charge consistently and submit to the same standards demanded of their teams, the message is loud and clear: safety is, in fact, a genuine value. However, when these same rules are relaxed because of haste, pressure for production or the easy way out, they lose their real meaning. In this context, the risk increases, because what once served to protect becomes just an empty formality, something that no one really takes seriously.

Setting the agenda and aligning priorities: Where is the focus?

The fact: The way a leader communicates, in what they say, what they repeat, what they measure and what they choose to spend their time on, clearly reveals what really matters. Even if they don't say it directly, they are always conveying messages that shape the team's outlook and choices.

In practice: Around 75% of employees adjust their own priorities based on what they perceive as valuable to their leaders. If safety is present in meetings, daily dialogues, appraisals and even recognition, it becomes understood as part of the essence of the workplace. But if it only appears on a one-off, protocol basis or only after an incident, it loses strength in relation to targets such as production, costs or deadlines.

What you say matters, but what you live in has even more impact. Safety needs to be at the center of the agenda, with coherence, presence and truth. According to the Harvard Business Communication Study, 2023.

Creating a climate for learning or guilt: The reaction to error.

The fact: The way a leader reacts to a mistake says a lot about the culture they are building. Their response can open space for learning or raise walls of silence and fear.

In practice: When the leadership chooses to see error as an opportunity, investigates causes without looking for culprits and commits to improving the system, the organization learns. Such environments have fewer work accidents and more evolution.

On the other hand, when mistakes are met with punishment, excessive demands or judgments, fear sets in. And fear silence, hides and holds us back. Safety disappears where there is no room to be human.

Cultivating psychological safety and engagement: The trust that sustains culture.

The fact: Leaders who create an environment where people feel free to speak up, ask questions, disagree and even make mistakes without fear build relationships of trust and solidity.

In practice: This kind of environment generates more engagement, well-being and cooperation. People stay, feel part of it and contribute more genuinely. Psychological safety is not an extra, it is the fertile soil where communication grows, risks are perceived in advance and problems are tackled before they escalate. It is where culture becomes alive, strong and, above all, human.

Direct correlation with overall performance: The example of high reliability organizations (HROs).

The fact: In organizations that achieve exceptional levels of safety, such as those that manage air traffic control, operate nuclear power plants or maintain large chemical industries, the difference doesn't lie in luck or cutting-edge technology. What sets them apart is an active, clear and constant commitment from top management. In more than 90% of these organizations, according to a recent study, the direct and sustained involvement of leaders is identified as an essential factor for their results (High Reliability Organizations Study, 2022).

In practice: What comes from the top profoundly shapes the culture. It is leadership that determines priorities, allocates adequate resources and ensures that safety is embedded in all strategic decisions, never being reduced to a simple department or operational cost to be controlled. In these organizations, safety is not just part of the conversation, it is the foundation that underpins every conversation.

There is no doubt about that. Leadership is not a peripheral detail, but the essential factor responsible for igniting or extinguishing the flame of cultural transformation.

Trying to change a safety culture without genuine leadership involvement is like trying to climb Everest from the wrong side of the mountain. It represents unsustainable effort and wastes energy. When the leadership decides to take real action, all the other elements naturally find their way. Without this, the risk is still there, and the culture insists on repeating what should never be accepted as normal.

VISIBLE AND COMMITTED LEADERSHIP: ACTION SPEAKS LOUDER THAN A THOUSAND WORDS

It is not enough to say that safety is important, you must demonstrate, consistently and truthfully, that this commitment is part of everyday life. This is what is known as "heartfelt leadership", which is perceived vividly and concretely by the team.

More than speeches or documents, what really builds culture are gestures that touch, attitudes that are repeated, decisions that confirm care. When leadership acts consistently, is present in detail, listens, observes, asks questions and really cares, it is felt.

And for this presence to become a reality daily, certain behaviors need to be part of the routine, with authenticity, not out of obligation. Because leadership that inspires is leadership that commits to what it teaches and lives what it believes in:

1. **Active and attentive presence in the field (purposeful visibility):** The Leader will see.

What: Being where work happens is a choice that speaks volumes. Getting out of the office, dedicating time and real attention to seeing the routine through the eyes of a caring leader. Not to watch, but to connect.

How: Not with the gaze of someone looking for mistakes, but with the intention of observing carefully, listening with genuine interest, talking about everyday challenges, recognizing good practices and showing genuine care for people's well-being.

Impact: The physical, respectful and curious presence of the leader strongly communicates how much safety really matters. Credibility is built by meeting, looking and listening.

Practical Tip: Include safety walks in your agenda, without the focus on auditing, but as a space for conversation and insight. Ask: "What concerns you about safety today?" or "What can we do to make your work safer?"

2. **Clear, frequent and two-way communication:** open dialog.

What: Create a flow of communication about safety that is constant, understandable, transparent and, above all, works both ways.

How: Reinforce expectations clearly, share learnings without pointing blame and ensure that there are accessible and reliable channels for people to express their concerns, report risks and suggest improvements with freedom and safety.

Impact: Strengthens alignment between everyone, increases trust and invites workers to become an active part of building solutions. When there is room to talk, there is more chance of acting before something happens.

Practical Advice: Are your communication channels open and secure? Do the people around you feel comfortable speaking up and know that they will be listened to with respect?

3. **Unshakable coherence between speech and action (walk the talk):** The test of truth.

What: Ensuring that every attitude and decision of the leader is in total alignment with what he says and the values he defends.

How: Avoid inconsistencies, empty speeches or contradictory practices. Having the courage to make difficult decisions that put safety above immediate pressure, such as stopping an operation, postponing a delivery or investing in more robust controls.

Impact: Nothing builds or destroys trust faster than consistency between word and deed. When example confirms speech, leadership is strengthened. When it doesn't, credibility crumbles.

Example: Attention to detail: if you demand the use of PPE, but enter the area without goggles, the message that comes through is not what you say, it is what you do. And that is worth more than any speech.

4. **Fair and Systemic Accountability:** Focus on improvement.

What: Build and maintain a clear system of accountability in safety, perceived as balanced, transparent and aimed at improvement, not just punishment.

How: The leadership fully recognizes its responsibility for systemic failures. When dealing with individual errors, it always adopts the principles of Just Culture, clearly distinguishing between unintentional errors, conscious risk behaviors and intentional violations. The approach remains continuously focused on learning and prevention, avoiding any attempt to seek blame or fuel accusations.

Impact: Creates a safe environment to talk about mistakes without fear. This fosters continuous learning, strengthens trust and allows for real progress in risk management.

Practical advice: Whenever you deal with a mistake, ask: "What in the system allowed this to happen?" This question opens the way to solutions, not just judgments.

5. **Strategic allocation of resources:** Where the money (and time) really goes.

What: Making the commitment to safety visible through the proper allocation of resources, not only financial, but also time and people.

How: Go beyond buying equipment. Ensure time for training, dialog, planning and investigations. Avoid overloading teams, recognizing that excessive demand compromises safe decisions and increases risks.

Impact: When the necessary resources are not made available, the transmitted message becomes quite clear, indicating that safety may be mentioned in speeches, but not actually experienced in practice. On the other hand, investing intentionally strengthens the culture, protects lives and sustains trust within the organization.

Practical advice: Periodically review where leadership time and the area's budget are being invested. What doesn't receive attention and resources doesn't sustain itself as value.

TRANSFORMATIONAL LEADERSHIP: ELEVATING COMMITMENT BEYOND COMPLIANCE

While Felt Leadership is an essential foundation for sustaining a consistent Safety Culture, reaching higher levels of maturity, especially the Independent and Interdependent stages, demands something additional. It requires leadership with transformative power, capable of inspiring not only by its constant presence, but above all by the purpose it carries.

We are talking here about Transformational Leadership, a style that goes beyond mere process management and target-based control. This approach touches people deeply, tapping into their inner motivations, personal values and authentic desire to participate in something bigger and more meaningful.

Widely studied in organizational psychology, especially by authors such as Bass and Avolio, Transformational Leadership promotes real changes, not only in immediate results, but also in the way people see their work and themselves. Rather than limiting itself to exchanges or rewards, this leadership raises the level of consciousness, broadens engagement and strengthens a genuine sense of belonging.

When applied to the world of safety, the four pillars of Transformational Leadership reveal a powerful path to establishing true connections and achieving sustainable change.

1. **Idealized influence (building trust and respect):** The leader as an ethical role model.

What: Act as a benchmark for ethical conduct and genuine commitment to the organization's values, especially safety. Demonstrate integrity, consistency and courage in every decision.

How to: Put care for the team and safety above momentary pressures or personal interests. Being the first to comply with the rules, especially the most critical ones. Taking responsibility when something doesn't go as expected, looking at the system before blaming.

Impact: Generates a bond of respect and trust that strengthens engagement. People begin to follow the leader not out of obligation, but because they feel inspired and represented by the example they set.

2. **Inspirational motivation (articulation of vision and purpose):** The leader as a communicator of meaning.

What: Conveying a clear, engaging and mobilizing vision of the future, an environment where safety is lived as a value and care between people is natural and constant.

How: Speak with enthusiasm and hope. Connect what each person does every day to the organization's greater purpose: protecting lives, caring for each other and building a safe and sustainable environment. Show that safety is meaningful, impactful and transformative.

Impact: Awakens a genuine desire to do more and better. It connects people emotionally to the cause of safety and inspires actions that go beyond what is necessary, driven by meaning and not just obligation.

Practical advice: Bring real stories into the conversation. Positive or learning stories help bring the purpose to life and show, in a concrete way, the value of each safe choice.

3. **Intellectual stimulation (encouraging innovation and critical thinking):** The leader as a catalyst for ideas.

What: Foster an environment where creativity, reflection and innovation are actively applied to safety. Stimulate thinking beyond the obvious.

How: Question what has always been done with a curious and constructive eye: "How can we do this more safely?". Encourage new approaches while respecting risk analysis. Create a safe space for ideas to be shared and mistakes to be treated as part of the learning process.

Impact: Awakens the creative potential of teams, generates more effective and sustainable solutions to safety challenges and strengthens a living culture that learns, evolves and improves continuously.

4. **Individualized consideration (mentoring, coaching and support):** The leader as a developer of people.

What: Taking a genuine interest in the growth, well-being and development of each person on the team, recognizing their individuality and potential.

How: Take the time to get to know everyone's stories, challenges and aspirations. Practice active listening, offer personalized feedback, acknowledge contributions sincerely and support the development of skills, including those related to safety.

Impact: Strengthens the connection between leader and led, increases engagement and awakens a sense of individual and collective responsibility for safety. People who feel cared for take more care of themselves, others and the environment.

In the specific context of Safety Culture, Transformational Leadership manifests itself concretely by:

Genuine and empathetic listening: Dedicating quality time to listening to experiences, fears, ideas and frustrations, welcoming without judgment.

Connecting safety, purpose and personal values: Helping each person to realize that safety goes beyond the rules of caring for one's own life, for that one loves, for one's colleagues, for the legacy one builds.

Motivation by example, care and recognition: Inspire with coherent actions, show real interest in people and value attitudes of mutual care more than targets met or punishments avoided.

Actively building trust and psychological safety: Creating an environment where it is possible to be vulnerable, admitting mistakes, raising your hand, asking for help and challenge what is not safe, without fear, with respect and a focus on learning.

ENGAGING SENIOR LEADERSHIP (C-LEVEL AND BOARD): THE STRATEGIC LANGUAGE OF SAFETY

Although leadership at all levels plays a decisive role, it is the authentic and continuous engagement of top leadership, such as the CEO, CFO, COO and Board of Directors, that offers legitimacy and real strength to Safety Culture. These executives set the strategic direction, allocate the most relevant resources, establish governance and make it clear to the entire organization that safety is not just lip service, but a core, non-negotiable value.

Without this visible sponsorship, cultural initiatives may get off the ground, but they are unlikely to reach depth and continuity.

Therefore, gaining the attentive eye and genuine commitment of these executives, who are constantly involved in multiple pressures and high-impact decisions, requires strategic intelligence from safety professionals and middle management. It is necessary to break away from an exclusively technical language, translating safety into topics that speak directly to the central concerns of senior executives, such as reputational risks, business stability, sustainability, solid governance and lasting results.

Presenting safety as an integral element of the organizational strategy, and not as an isolated topic, is the most solid way to create connection, engagement and ensure that it is effectively an absolute priority at the top.

1. **Explicit connection with business strategy and objectives:** safety as an enabler.

What: To show that safety is not just a cost center, but a strategic element that drives the organization's larger objectives.

How: Relate safety directly to areas that matter to the business:

Operational efficiency: Less downtime, less time off work, more productivity.

Reputation and brand: Strengthened image in the eyes of clients, investors and society.

Sustainability and ESG: Compliance with environmental, social and governance commitments, and obtaining a social license to Operate (LTO).

Attracting/retaining talent: Safe environments are magnets for the best professionals.

Shareholder value: Less work accident costs, less exposure to legal risks and greater market confidence.

Impact: Repositions safety as an essential part of business success, integrated into the strategy, not as an operational appendix, but as a solid basis for growing responsibly.

2. **Focus on material and high potential risks (SIFs):** What really matters to the top.

What: Directing communication towards low-frequency risks with severe consequences such as serious injuries and fatalities (SIFs)

How: Present objective data and clear analysis of critical risks: potential scenarios, financial impact, repercussions on the company's image, legal implications and threats to business continuity. Using visual tools such as Bow-Tie can help make this information more accessible and strategic.

Impact: Attracts the focus of top management to what can really compromise the organization, shifting attention from superficial indicators to what is really vital.

3. **Financial language and enterprise risk management:** Speaking the language of the C-Level.

What: Converting safety issues into the language used in the strategic and financial decisions of top management.

How: Present the total costs of work accidents, direct, indirect and opportunity costs. Demonstrate the ROI (return on investment) of safety actions. Frame safety within the Enterprise Risk Management (ERM) model, using terms such as impact on EBITDA (earnings before interest, taxes, depreciation and amortization), Value at Risk (VaR) and cost-benefit analysis.

Impact: Repositions the safety professional as a strategic ally of senior management, facilitating C-Level engagement and prioritization of resources for preventive initiatives.

4. **Strategic benchmarking and best practices:** Learning from leaders.

What: Compare the organization's safety performance and practices with those of companies in the same sector, direct competitors or recognized benchmarks.

How: Use benchmarking data to contextualize where the organization is, highlight gaps and awaken a sense of urgency or opportunity for improvement. Show where you can get to with well-directed strategic decisions.

Impact: It demonstrates that safety is not just a differentiator, but an intelligent and consolidated management practice among leading companies, which can inspire action and strengthen the engagement of top leadership.

5. **Clear governance and accountability at the top:** Defining roles and expectations.

What: Suggest and support the creation of a well-defined governance structure for safety, with clear roles for top leadership.

How: Establish formal C-Level and Board responsibilities for safety oversight. Create relevant KPIs, both proactive and reactive, aimed at this strategic level. Include safety on regular agendas and integrate its performance into executive targets and remuneration.

Impact: Ensures that safety receives the level of attention and priority it deserves, being treated as part of excellence management, with clear responsibility at the highest level of the organization.

6. **Concise, visual and action-oriented communication:** Respecting the executive's time.

What: Present safety information in a direct, visual way that focuses on the strategic decisions that need to be made.

How: Use clear dashboards, trend graphs and objective executive summaries.

Avoid excessive use of technical jargon and lengthy reports. Highlight the key points: relevant trends, emerging risks, recommended actions and resources required.

Impact: It makes it easier for senior management to read, understand and make decisions, making the most of the time available and increasing the effectiveness of safety interactions.

The engagement of top management doesn't happen in a single moment, it is built up little by little, in a continuous journey of connection, mutual learning and trust. it is a process that requires presence, listening, clear communication and, above all, the constant demonstration that a strong Safety Culture is not only a human value, but also a strategic asset for the business.

For this to happen, safety professionals and middle managers need to expand their skills. They need to look at the whole picture, understand the impact of safety on strategy, speak the language of business clearly and develop the ability to influence with consistency and purpose.

Because, in the end, it is not just about convincing. It is about building a path together where safety, performance and care go hand in hand.

CONCLUSION: LEADERSHIP AS A FOUNDATION, COMPASS AND VITAL ENERGY

Leadership is not just another element in Safety Culture, but the very foundation on which everything is built. It is the firm foundation that supports the precise compass that guides and the vital force that drives the organization towards a more mature, conscious and humane culture.

Leaders at all levels carry a responsibility that cannot be transferred. It is they who influence the beliefs, values and practices of their teams daily through their choices, words and attitudes. It is in the daily routine, in everyday conversations, in difficult decisions and repeated gestures that shape behavior and consolidate values.

When these leaders choose to act with genuine and authentic presence (Felt Leadership), when they inspire and develop their teams through the principles of Transformational Leadership, and when they clearly communicate the value that security has for both the business and people's lives, they create a fertile environment in which people and processes evolve in harmony. And when they clearly communicate the value that safety has for both business and people's lives, they create a fertile environment in which people and processes evolve in harmony.

More than managing safety, these leaders drive it with purpose, consistency and sensitivity. They are the ones who turn care into culture, and culture into legacy. After all, true excellence is not measured only by numerical indicators, but by the genuine commitment to life that underpins every result achieved.

PRACTICAL ADVICE:

Be seen: Your presence in the field, listening and observing, is worth a thousand memos.

Ask "Why?" and "How can I help?": Show curiosity and a willingness to remove barriers.

Connect the dots: Articulate how safety drives business results (efficiency, reputation, talent).

Invest in capabilities, not just compliance: Focus on building resilience and competence.

Demand results but celebrate the process: Recognize proactive efforts and learning, not just the absence of work accidents.

QUESTIONS FOR REFLECTION:

1. As a leader, do my daily actions consistently demonstrate that safety is a non-negotiable value for me?
2. Does my team perceive my commitment to safety as genuine (Felt Leadership)?

3. What elements of transformational leadership (inspiration, challenge, support) can I strengthen in my practice?
4. How can I best communicate the strategic value of safety to my superiors and peers?
5. Am I creating an environment where people feel safe talking about risks and mistakes (psychological safety)?

9

NAVIGATING THE JOURNEY

METRICS AND INDICATORS THAT LIGHT THE WAY

In any journey of transformation, whether it is climbing a mountain or building a new culture, it is essential to know exactly where we are, where we want to go and whether we are, in fact, heading in the right direction. In the context of Safety Culture, indicators and metrics play this fundamental role, acting as a compass and a map.

They provide clarity about our current situation, help us make strategic decisions and allow us to assess whether the path we are on is generating the desired results.

However, just as important as measuring is knowing exactly what to measure and how to interpret this data. Poorly chosen or misunderstood metrics can create a false sense of safety, distort priorities or even encourage behavior that goes against the culture we want to strengthen.

Therefore, more valuable than cold numbers are metrics guided by a clear purpose, capable of honestly revealing where we really are, inspiring conscious decisions and keeping alive the deepest essence of safety, which is true care for life.

THE SEDUCTIVE TRAP OF REACTIVE INDICATORS (LAGGING INDICATORS)

Historically, safety management has been guided almost exclusively by reactive metrics, the so-called lagging indicators. These record what has already happened: work accidents, losses, failures. Among the best known are:

- Lost Time Accident Frequency Rate (TFCA or LTIR) Accident Severity Rate (Number of days lost).
- Total number of recordable work accidents (TRIR - Total Recordable Incident Rate).
- Number of fatalities.
- Costs associated with work accidents (direct and indirect).
- Number of non-conformities in audits.

These indicators are undoubtedly important. They reveal the real impact of past events and are often demanded by regulatory bodies, investors and clients.

However, when used as the sole benchmark for assessing safety and your organizational culture, they carry a silent risk, creating a false impression that if the numbers are low, everything is under control and safe.

And that is where the danger lies. Because Reactive Indicators only look in the rearview mirror. They show the consequence, but don't reveal the cause. They don't anticipate, they don't prevent, they don't help you see the signs before the worst happens.

Relying solely on them to make decisions or measure culture can lead to misleading conclusions and, worse, generate behaviors aimed at "avoiding the number" rather than transforming reality.

True maturity lies in balancing our gaze, learning from the past, yes, but acting in the present with a focus on the future.

1. **Focus on the past, not prevention:** Reactive Indicators look at what has already happened. They record failure, but don't capture the ability to prevent it. Using only these metrics to guide safety is like trying to drive looking only in the rear-view mirror: driving becomes risky and reactive, never anticipatory.

2. **Low frequency (especially for serious events):** Serious work accidents and fatalities are fortunately rare. And this, while positive, also means less data available for analysis over short periods. The absence of these events can create the dangerous illusion that everything is fine, when in fact serious risks are still present and just haven't materialized yet.

3. **They don't measure the strength of skills:** A good number doesn't always reveal a good culture. Low work accident rates can be the result of luck, under-reporting or lower exposure, and not necessarily of effective systems. These indicators do not show the level of engagement of teams, the maturity of leadership or the resilience of the organization in the face of unforeseen situations.

4. **They can encourage unwanted behavior (perverse effect):** The blind pursuit of targets such as "zero work accidents", when based only on final figures, can lead to the omission of near misses, the under-recording of injuries or pressure to hide failings. This creates a culture of fear and make-up, where the numbers look good but the reality remains vulnerable.

5. **Lack of sensitivity to cultural changes:** Transformations in culture take time, and reactive indicators are slow to reflect these changes. This can demotivate teams or lead to erroneous assessments, such as imagining that nothing has changed when, in practice, a lot is already being built. Because they are the "end of the line", this data doesn't always tell the full story.

Reactive Indicators tell us what went wrong yesterday, but they don't tell us how to prevent the same problem from happening tomorrow. They are essential for recording events, accountability and understanding the past, but insufficient for guiding the future.

When used in isolation as the main management tool, these indicators can create a false sense of control, hide weaknesses and even distract them from what really matters, which is preventing risks, learning from each situation and continuously evolving.

These indicators have their relevance, of course, but they shouldn't take center stage in management. Building a strong and sustainable Safety Culture requires more than looking backwards, it requires a clear, intentional and careful vision towards the future.

THE POWER OF PROACTIVE INDICATORS (EFFORT AND CAPACITY/ LEADING INDICATORS)

To move forward with clarity and safety on the journey of cultural transformation, it is essential to go beyond indicators that only reveal what has already happened. Proactive indicators, also known as leading indicators, must be complemented and often prioritized.

These indicators work like the dashboard of a car or the flight instruments of an aircraft, offering real-time signals about what is happening behind the scenes. They reveal what is being done, identify what still needs to be adjusted and flag up possible risks before they become real.

These metrics evaluate efforts, conditions, behaviors and capabilities that directly precede incidents. They focus on the processes and everyday elements that can strengthen or weaken safety.

Rather than just recording losses, these indicators help us to build safe paths, enabling conscious decisions, agile adjustments and, above all, concrete actions that reinforce both prevention and Safety Culture.

WHY ARE THEY POWERFUL?

Focus on prevention: They look ahead, allowing action to be taken before accidents occur.

Visibility of root causes: Helps to see what really influences the results, whether they are factors that strengthen or weaken safety.

They measure capabilities: They assess the quality of leadership, the maturity of processes, the preparedness of people and the consistency of culture.

Engagement and action: Because they are more tangible and closer to operational reality, they facilitate direct action and mobilize all levels of the organization.

Sensitivity to change: They pick up signs of evolution or regression more quickly than Reactive Indicators and are more agile thermometers of culture and processes.

Characteristics of good Proactive Indicators:

Measurable: Allows clear monitoring, with objective quantitative or qualitative data.

Actionable: They generate useful information that guides practical decisions and specific interventions.

Predictive or correlated: There is logic or evidence that improving these indicators contributes to reducing work accidents and strengthening the culture.

Process-focused: They measure the consistency of the system, not just the one-off performance of activities.

Engaging: They are understandable by everyone, relevant to everyday life and encourage active participation.

Contextualized: Aligned with the real risks, the maturity of the culture and the specific scenario of the organization.

Categorizing Proactive Indicators: The PDCA cycle as a structured guide. There is a wide range of potential Proactive Indicators. To organize them logically and ensure comprehensive coverage, we can use the Plan, Do, Check, Act (PDCA) continuous improvement cycle as a framework:

(P) Plan: Measuring preparation and strategy.

THEY ASSESS THE QUALITY OF PLANNING AND PREPARATION TO MANAGE RISKS AND BUILD CULTURE.

Examples:

Quality of risk analysis: Percentage of analyses (APR, HAZOP) completed with satisfactory quality (clear criteria, front-line participation).

Culture maturity level: Result of periodic diagnostics (surveys, interviews) compared to a model.

% of safety budget implemented: Monitoring the application of planned resources.

No. of critical risks (SIFs) identified with action plans: Focus on the risks with the greatest potential.

Legal/standard compliance index: Percentage of requirements met (verified by audits).

(D) Do: Measuring implementation and execution

THEY MONITOR THE EXECUTION OF PLANNED ACTIVITIES AND THE IMPLEMENTATION OF BARRIERS ON A DAY-TO-DAY BASIS.

Examples:

Quality of safety interactions: evaluation of the quality (and not just the quantity) of Daily **Safety Talks**, behavioral dialogues and inspections carried out by leadership, with a focus on constructive feedback and the identification of real barriers.

Training effectiveness: percentage of training completed and practical evaluation of the application of knowledge in everyday life (through observations or tests).

Voluntary reporting rate: number of records of near misses, unsafe conditions and good practices. An increase can reflect trust and engagement.

% of safety initiatives implemented: monitoring the progress of planned safety projects and actions.

Compliance with critical processes: percentage of audits carried out on processes such as Work Permits (PTW), LOTO, among others, within the expected standard.

(C) Check: Measuring verification and monitoring

THEY EVALUATE THE EFFECTIVENESS OF CHECKS, AUDITS AND CONTINUOUS MONITORING.

Examples:

Percentage of action plans completed on time: Rate of closure of actions arising from audits, investigations or evaluations.

Quality of audits/inspections: Evaluation of the depth, relevance of the findings and effectiveness of the actions generated.

Frequency and quality of leadership routines: Percentage of planned routines (such as safety walks) and analysis of the quality of these interactions.

Employee perception (pulse): Results of quick, targeted surveys on topics such as trust, freedom to report, perception of listening and welcome.

(A) Act: Measuring learning and continuous improvement.

They focus on the organization's ability to learn from experience and transform these lessons into concrete actions for systemic improvement.

Examples:

Number of lessons learned shared and implemented: Monitoring how lessons learned are disseminated and effectively applied daily.

Average time for investigation and implementation of actions: Assessment of the organization's agility in responding to events and turning failures into real improvements.

Number of improvements implemented via suggestions: Measurement of the contribution of the front line to the continuous improvement of safety processes.

Benchmarking results: Comparative use with other organizations or units to identify opportunities for improvement.

Frequency of review of policies/procedures: Analysis of the periodicity and consistency in updating guidelines based on accumulated learning.

IMPLEMENTING AN EFFECTIVE INDICATOR SYSTEM: BEYOND DATA COLLECTION

Developing a good set of Proactive Indicators is only the beginning. The real value comes when these indicators are integrated into a living management system, capable of using them to drive real, consistent and sustainable improvements:

Careful and participatory selection: Less is more. Choose a few indicators, but ones that are relevant to critical risks, aligned with the strategy and compatible with the organization's level of maturity. Involving different levels in this process ensures engagement and ownership.

Clear and simple definition: Each indicator needs to have a clearly defined purpose, specifying precisely what will be measured, how this measurement will be carried out, who will be responsible for collecting and monitoring the information and how often this will happen. Being clear on these aspects avoids misinterpretations and promotes greater consistency and effectiveness in safety management.

Smart goals (SMART), focus on the process: Set goals that are challenging but also realistic, specific and measurable. It is essential to understand that the goal is not just to reach a certain number, but to improve the processes represented by it. Be careful not to create targets that could pressure people into manipulating data, thus compromising the truth and effectiveness of actions.

Transparent and visual communication: Indicators must be communicated in an accessible way, with simple and objective dashboards. More than just showing numbers, it is important to put them into context, allowing everyone to understand what's behind each result.

Critical analysis and dialog: Indicators are not a tool for judgment, but a starting point for productive conversations. They should serve as a trigger for reflection, learning and adjustment. Meetings with leaders, committees and teams should discuss variations and identify ways to improve.

Connection with Accountability: Integrate Proactive Indicators into performance evaluations in a fair, balanced and constructive way. The focus should be on continuous effort and progress, not just the final figure.

Continuous review and adaptation: A good indicator system is dynamic. It should be reviewed frequently, keeping pace with changes in culture, processes and the business context.

By balancing the focus between Reactive Indicators and a well-structured set of Proactive Indicators, the organization can see the true health of its Safety Culture in greater depth and in advance.

What was once just a rear-view mirror is now a beacon illuminating the path of transformation, strengthening capacities and pointing clearly to a safer, more conscious and more humane future.

PRACTICAL ADVICE:

Question "Zero Work Accidents": Go beyond the number. Ask which Proactive Indicators are supporting this goal and how the organization is preventing underreporting.

Prioritize Proactive Indicators: Request reports that show prevention efforts, the consistency of practices and the health of the culture, not just accident rates.

Invest in data quality: The right decisions depend on reliable information. Encourage in-depth and contextualized analyses, avoiding superficial readings.

Use data to learn, not to punish: Create an environment where indicators are triggers for constructive conversations, not tools for pressure or punishment.

Connect metrics to strategy: Make sure that safety indicators are integrated into the larger business objectives, because safety is also strategy.

QUESTIONS FOR REFLECTION:

1. What indicators (Reactive and Proactive) does my organization use today? Do they really show us the quality of our safety or just what has already happened?

2. Do our Proactive Indicators assess the critical capabilities and processes that underpin safety, or are they only focused on one-off, low-complexity tasks?

3. How are we using indicator data? Do they serve as a source of learning and continuous improvement, or are they mainly used for reporting and judgment?

4. Could the way we deal with targets for Reactive Indicators be, even unintentionally, encouraging under-reporting in our organization?

5. What new Proactive Indicators could we adopt to get a deeper reading of our real risks and the maturity of our culture?

10 LEARNING FROM EXPERIENCE

THE VITAL ART OF RESEARCH AND DEEP ORGANIZATIONAL LEARNING

Theory about Safety Culture, transformative leadership, empowering processes and genuine engagement comes to life when it is filtered through real experience. It is in the day-to-day life of organizations, among discrete successes, small deviations, near misses and unsafe conditions, that everything takes on meaning and, more than that, becomes applicable in practice.

Learning from what happens every day, and not just from major tragedies that make the headlines, is perhaps one of the most valuable skills for any organization that wants to go beyond simple continuous improvement and effectively build a resilient culture.

But it is essential to remember that real learning doesn't happen spontaneously. It doesn't just come about with time or repetition. It requires a clear and constant intention.

In-depth, well-structured and coherent investigative processes must be guaranteed. It is necessary to cultivate a safe environment where people can express the truth, raise important issues, admit mistakes or report situations that nearly end in work accidents. Above all, it requires committed leadership that listens attentively, welcomes with respect, acts quickly and turns lessons learned into real actions that are monitored, reviewed and integrated into the routine.

Only in this way can learning stop being just theory and become real culture. Learning is something that is measured by the concrete changes it generates. Where there are no visible changes, there is no real evidence that learning has taken place.

BREAKING WITH THE TOXIC TRADITION OF GUILT: TOWARDS A SYSTEMIC UNDERSTANDING

For too long, the reaction to an incident in many organizations has been to quickly find out "who did it wrong". The response, which is closer to a witch-hunt than a search for understanding, falls on the individual, almost always someone on the front line, who must carry the weight of what happened alone.

The logic is simple but shallow: if something has gone wrong, someone must be blamed. And from this, "solutions" emerge that, in practice, solve little:

Retraining: The same procedure that has already proved insufficient, as if repeating the instruction were enough to avoid failure.

Punishment: Warnings and even dismissals, which instill fear rather than care.

More bureaucracy: New rules, extra checklists, barriers that hinder real work, without solving the root of the problem.

This way of dealing with failures, focused on finger-pointing, is deeply damaging. It blocks listening, silences reporting and prevents learning. The culture of fear doesn't teach. It only hides.

If we want to evolve, we need to replace judgment with curiosity, and isolated punishment with a systemic analysis. Because real safety is not built by blaming people, but by understanding contexts.

1. **It silences the front line:** Fear paralyzes. When there is fear of punishment, exposure or humiliation, people keep quiet. They stop reporting near misses, failures and unsafe conditions. And without this information, the so-called "weak signals", the organization loses the chance to act before something serious happens. Silence creates a dangerous cycle: less data, more risks, more punishment... and even less information.

2. **It blinds us to the real causes:** Looking only at the mistakes of those in the operation diverts attention from what really created the risk scenario. Poor design, inadequate tools, insufficient training, excessive pressure, communication failures, management decisions - all of these are usually behind it. When we ignore these systemic causes, the mistake is repeated. The only thing that changes is the name of the culprit.

3. **It creates a false sense of safety:** Applying a punishment or creating a new rule may seem like action, but it is an illusion. The problem is still there, only now hidden under the apparent sense of control. When leadership is satisfied with palliative measures, vigilance decreases and the risk remains, camouflaged and ready to emerge again.

4. **It destroys trust and psychological safety:** A culture based on blame weakens bonds and creates barriers between leaders and teams. Without psychological safety, without the certainty of being able to speak honestly, admit fault or ask for help, learning dies at its roots. Collaboration weakens, care is hidden, and continuous improvement becomes unfeasible.

5. **It feeds hindsight bias:** After something happens, everything seems obvious. "How could you not have seen that? But this judgment ignores the real context in which the decisions were made in a hurry, with partial data, under pressure. Understanding what really happened requires empathy and listening. A mature investigation seeks to understand the local logic, why that choice made sense to those who were there at the time. Because this is the only way to truly learn.

THE ALTERNATIVE: EMBRACING SYSTEMIC UNDERSTANDING

A mature and effective approach, aligned with a fair and resilient culture, abandons the search for culprits and turns to a deep understanding of the system. The aim is not to point fingers, but to understand, truthfully and sensitively, why that event happened, considering the complex interaction of factors that make up the real working environment. This analysis involves looking closely at

People: Their skills, limitations, physical and emotional state now.

Technology and equipment: The design, maintenance, usability and actual operating conditions.

Processes and procedures: Whether they are clear, adequate and feasible in practice.

Organizational environment: The prevailing culture, the presence (or absence) of leadership, invisible pressures, internal communication and available resources.

THE INVESTIGATION BEGINS TO SEEK ANSWERS TO DEEPER QUESTIONS:

- What were the real conditions and pressures at the time of the event?
- Were the procedures adequate to the operational reality?
- Were the tools and training offered sufficient and effective?
- Did communication work as it should have?
- Have previous management decisions contributed to exposing vulnerabilities?

- Why, in that context, did the actions taken seem reasonable to those who carried them out?

- Which defenses failed and, just as importantly, which barriers worked and prevented further damage?

Adopting this systemic vision does not eliminate the importance of individual responsibility. Situations of gross negligence, intentional violations or malicious conduct still require appropriate accountability. However, this accountability is applied fairly, carefully considering the context and ensuring balance.

In this scenario, the concept of accountability takes on a broader dimension. It does not fall exclusively on those who carry out a task, but also on those who lead, who design the systems and who create the organizational environment. The real commitment is to ensure that doing the job safely doesn't just depend on people's isolated efforts, but is a viable, sustainable choice that is naturally integrated into the organization.

THE RESEARCH PROCESS: TOWARDS RIGOR, DEPTH AND AUTHENTIC PARTICIPATION

For the investigation to really generate learning, and not just another report, it needs to go beyond formality. The process must be structured intentionally, conducted in depth and, above all, built with the participation of those who know the reality of the work as it is, and not as it is described on paper:

Intelligent selection (where to focus on the magnifying glass?): Not every event requires an in-depth investigation. Having clear criteria helps direct energy to what matters most. The focus should be on events with the greatest potential for learning or risk, such as high potential near misses (SIF Potentials), relevant deviations and recurring situations that reveal weaknesses in the system. The idea is to learn before the worst happens.

Multidisciplinary teams (different perspectives): A good investigation cannot be based on just one point of view. It is essential to bring together different perspectives and include diverse voices in the process, such as operators, supervisors, engineers, safety professionals and human factors experts. In addition, involving someone from outside the area or a neutral facilitator can significantly enrich the investigation, bringing objectivity and questioning assumptions that have often crystallized into habits that are difficult to perceive.

Comprehensive data collection (beyond the obvious): The quality of the investigation is directly linked to the quality of the data collected. This requires going beyond the first report. You need to examine the site, observe physical evidence, review records (such as logs and maintenance), analyze documents (procedures, risk analyses, training) and, above all, conduct interviews with attentive listening.

These conversations should take place in a safe environment, where those who took part in the event feel welcomed and respected. The aim is not to judge, but to understand, with empathy, why the decisions made sense at that time, in that context.

Because it is in this guilt-free listening that the learning that transforms is born.

Deep causal analysis (peeling the onion): Digging deep transcends simply identifying what happened, as it involves understanding why the system allowed that to happen. It is not enough to attribute failures to "human error", you must go further and uncover the contributing factors, revealing the root causes hidden in the system. Tools such as the "5 Whys", the Ishikawa Diagram, the Fault Tree Analysis (FTA) and methods such as Causal Factor Analysis all have their value, but no tool beats a curious, open-minded team with a systemic vision. Above all, we seek a clear understanding of where and how the system failed in order to make it more robust and resilient.

Systemic and prioritized recommendations (actions that work): Recommendations should go beyond one-off adjustments or generic guidance, seeking to strengthen the barriers of the system, rather than just correcting individual behaviors. it is essential to prioritize robust actions, always respecting the hierarchy of controls, which begins with the complete elimination of risk, followed by substitution, engineering solutions, administrative controls and, finally, the use of Personal Protective Equipment (PPE). The higher the action is in this hierarchy, the more effective and long-lasting the proposed solution will be.

In addition, actions must follow clear criteria, be specific, measurable, achievable, relevant and time-bound, following the SMART model. Above all, it is essential to prioritize each measure by its real impact on effectively reducing the risks involved.

Clear and actionable report (communicating for change): The final report must be more than a document. It needs to clearly communicate the facts, the causal analysis and, above all, the recommended actions. No excessive technical jargon or texts that are difficult to translate into practice. The focus should be on learning and action. The reader needs to understand what happened, what needs to be done and why.

Sharing lessons learned widely (multiplying knowledge): Learning that is not shared is lost. Lessons need to be taken throughout the organization and, when it makes sense, even outside of it. Each area, each team, each person must receive information in a way that is adapted to their reality, because preventing mistakes from being repeated starts with ensuring that everyone learns from them.

Close monitoring of implementation (closing the loop): Investigating, recommending and not implementing is the same as ignoring the problem. It is essential that each action has a person responsible, a deadline and a follow-up process. Monitoring progress, checking that the actions have had the expected effect and adjusting what is necessary closes the learning cycle. This is the only way that change stops being an intention and becomes a practice.

Broad and transparent sharing of lessons learned: The true value of an investigation doesn't end at the report. It multiplies when the lessons learned are shared openly, carefully and intentionally with the entire organization. And, where possible and appropriate, also with other companies in the sector or the safety community.

It is not about exposing names or trying to hold individuals responsible. The real aim is to disseminate what matters most, which is knowledge about the system's vulnerabilities, the defenses that failed or proved effective, and the good practices that can prevent the same mistake from being repeated in other contexts, with other people.

For this learning to spread powerfully, it is essential to use various communication channels, such as team meetings, internal newsletters, short videos or anonymized case studies, and to adjust the language according to the audience. When a message reaches people in an accessible and respectful way, it has an impact. And when learning is transformed into collective action, the culture is strengthened.

LEARNING FROM EVERYDAY SUCCESS: THE HIDDEN WEALTH OF ORDINARY WORK

Limiting learning only to negative events, such as incidents, failures and deviations, is seeing only half the story. Real, everyday work is far from being a perfect execution of procedures. It is dynamic, full of variability, unforeseen events and adjustments that people make to get the job done, achieve goals and, most of the time, ensure safety even in the face of system imperfections.

Studying how this day-to-day success is built is a powerful and still little-explored source of learning. But how do you do it in practice?

Observe the work as it is done (Work-as-Done): Go beyond what is prescribed in the documents (Work-as-Prescribed). Be present at the Gemba with active listening and respect. The focus is not on inspecting, but on understanding. Notice the nuances, the adjustments, the responsibly created shortcuts, the things that really support safe working daily.

Talk about practice (not just problems): Talk to the teams about what works well. Hear about the solutions that have emerged in the field, the useful tools, the moments when they overcame a challenge. Use approaches such as Appreciative Inquiry to highlight what strengthens, what inspires, what works.

Analyze tasks with a focus on success: Use tools like Task Analysis to map out not only the risks, but also the strategies, skills and conditions that make work possible and safe. Understanding what contributes to good performance is just as important as understanding what leads to mistakes.

Share good practices: Create formal or informal spaces in which these solutions can be shared and multiplied, such as databases, conversation circles and communities of practice. Sharing knowledge is a genuine way of recognizing the value of those who carry out the work. Celebrating these contributions reinforces the idea that safety is also built through protagonism.

By learning from successes, we stop simply correcting failures and start actively strengthening what is already working well. In this movement, we broaden engagement, give a voice to the people on the front line and recognize their role as essential and active participants in building safer and more effective solutions.

CASE STUDIES AS COLLECTIVE LEARNING TOOLS: DRAWING UNIVERSAL LESSONS

In-depth analysis of case studies, whether historical work accidents, industry events or relevant internal situations, is a powerful tool for turning concepts into practice and generating real learning.

More than simply reporting what happened, the real value lies in the ability to identify systemic patterns, recognize blind spots and extract lessons that are applicable in different contexts. To do this, it is essential to avoid the trap of individual blame or details that distract attention from what is essential. Anonymizing or generalizing cases can be useful to keep the focus on what really matters, which are the lessons the incident teaches us.

DISCUSSING THESE CASES COLLECTIVELY, WITH LISTENING AND REFLECTION, HELPS TO DEEPEN UNDERSTANDING OF CRITICAL ISSUES SUCH AS:

Normalization of deviation: Small deviations that, when tolerated, erode defenses until the unacceptable becomes routine.

Change management failures (MOC): Changes made without proper analysis and planning, which introduce silent risks.

Safety vs. production dilemma: When poorly communicated or ambiguous goals put people in front of risky choices.

Safety **Culture vs. Overconfidence:** Lack of listening, low psychological safety or complacency about risks.

The **decisive role of leadership:** The direct impact of leadership decisions or omissions in building or weakening culture.

Principles of high reliability organizations (HROs): Constant concern about failures, sensitivity to operation, resilience in the face of the unexpected, valuing technical expertise and reluctance to simplify the complex.

Facilitating these conversations, connecting the lessons learned to the organization's reality, opens up space for reflection, questioning assumptions and strengthening the commitment to safety. It is in this process of exchange, listening and analysis that knowledge is transformed into collective awareness and culture gains consistency.

FROM LEARNING TO CONCRETE ACTION: CLOSING THE CYCLE OF CONTINUOUS IMPROVEMENT

Identifying lessons learned is just the starting point. The real challenge where many organizations stumble is in turning these lessons into concrete actions, implementing them consistently and sustaining them over time until they become a living part of the way the organization operates.

A well-written report or an inspiring workshop may thrill in the moment, but if they are not followed up by visible and real changes, they become just another noise. And more than that, they generate weariness, disbelief and cynicism. Because people watch. And when they see that nothing changes, they learn that the talk is nice, but the system remains the same.

Real learning requires action. It requires the courage to change routes, review decisions, take responsibility and rewrite practices. This is where culture is transformed, not in what is said, but in what is done with what is learned.

TO CLOSE THE LOOP AND GUARANTEE REAL IMPROVEMENTS:

Visible Leadership Commitment: Transformation starts at the top. The top leadership needs to be more than a supporter, it needs to be an active sponsor of the implementation of the lessons learned. This means allocating resources, removing obstacles and closely monitoring progress, showing with attitudes that safety is a real priority, not just talk.

Robust Tracking System: It is essential to have a formal system for monitoring the progress of actions. Who is responsible? What is the deadline? Has the action been completed? More than that: did it work? Did it solve the problem without creating new risks? Effectiveness needs to be measured, not just execution.

Transparent communication: People need to know what is going on. Sharing progress, the challenges faced, and the results achieved strengthens engagement and trust. Recognizing progress and celebrating lessons learned is also part of the process of strengthening the culture.

Integration into Management Systems: Lessons learned cannot be left unattended. They must be incorporated into policies, procedures, training, risk analysis and indicators. This is how you create organizational memory and prevent the same mistakes from being repeated in the future.

Culture of feedback and continuous review: Learning is not an end point; it is a cycle. Creating an environment that welcomes feedback on what has (or haven't) worked and allows for continuous review, adaptation and improvement is essential to sustaining any change.

The ability to learn in depth, both from mistakes and successes, and to transform this learning into concrete and consistent action, is what differentiates reactive organizations from those that consciously choose to shape a more secure, resilient and humane future. This doesn't happen by chance. It requires structured processes, a culture that listens and a leadership that acts.

CASE STUDY 1: THE "MODEL" FACTORY AND THE DANGEROUS ILLUSION OF CONFORMITY

THE SCENARIO:

Metalúrgica Progresso was seen as a benchmark. The numbers were low LTIR almost zero. Audits praised the processes. Paperwork was always up to date. The management proudly celebrated what they called a "mature safety culture".

Everything seemed to work perfectly. The image was excellent.

But, as is so often the case, the glossy surface hid deep cracks. Because culture is not only measured by statistics. It is revealed in the silences, the improvisations, the day-to-day choices and, above all, in what isn't in the reports.

THE HIDDEN REALITY (DISCOVERED BY A NEW EHS MANAGER):

Paper procedures: Long, complex, outdated, written to comply with requirements, not to provide real guidance. In practice, workers followed workarounds, which were considered more agile, but not always safe. The procedures only came to life on audit days. They were, as they said in the corridors, "for English to see".

Silence about risks: Near misses and deviations were rarely reported. Not for lack of awareness, but for fear. There was subtle fear in the air. The leadership, focused on indicators, especially the LTIR, didn't encourage frank conversations. A culture of silence took hold, discreet and dangerous.

Cosmetic audits: The focus of the audits was visual and documentary. Performance was measured by the appearance of the organization that day. The areas were "made up" to welcome the auditors. No one asked what it was like to work there on an ordinary Tuesday, under pressure and with tight deadlines.

Distant leadership: The board celebrated the figures but had little presence on the shop floor. There was rarely any real listening to day-to-day difficulties. Communication was top-down, full of rules but empty of connection.

Ineffective training: Theoretical, passive sessions with little room for participation or practical application. People "watched" the content but didn't see themselves in it. There was a lack of connection, purpose and retention.

Diagnosis: Metalúrgica Progresso did not have a mature culture. It had the illusion of safety compliance, well presented on paper, but with its real capabilities wearing away in silence. A system that appeared strong on the outside, but which, on the inside, was already showing signs of fragility. And no one dared to tell.

THE PAINFUL AWAKENING:

A serious non-fatal but high potential (SIF Potential) incident during a power outage exposed what many already felt, but few verbalized. The initial investigation attributed the cause to "human error". But it was the arrival of a new manager, with a systemic eye and active listening, that brought the truth to light: a confused procedure, pressure for production, ignored near misses and purely theoretical training formed the backdrop to that occurrence.

THE KEY TURN

1. **Useful procedures:** We carried out a complete review of the procedures, with the direct participation of the workers themselves. We prioritized clarity, applicability and practicality in everyday life. We simplified the documents, incorporated flowcharts, illustrative images and adopted accessible language. The sole and essential purpose was to make the procedures truly useful.

2. **Active reporting culture:** The leadership has firmly communicated that all reports are welcome, without punishment. Near misses, deviations, unsafe conditions were all valued as sources of learning. Channels were simplified and each report was given visible feedback, showing that listening was real.

3. **Proactive metrics:** LTIR is no longer the only thermometer used. Other indicators began to be monitored, including the number and quality of safety reports, the percentage of procedures reviewed together with workers, the hours dedicated by leadership to direct contact with the operation at the Gemba and behavioral audits focused on daily practice. In this way, the indicators began to more clearly reflect the culture of the organization, and not just the result.

4. **Leadership at the Gemba:** Directors and managers were encouraged and prepared to go out into the field in a structured way, with listening, presence and dialog. The Gemba ceased to be a place of passage and became a place of learning.

5. **Training that transforms:** Theoretical training has given way to the development of real skills. Simulations, case studies and practical assessments became part of the routine. Learning became more experienced.

6. **Investigations that teach:** Analyses are no longer looking for culprits. They started looking for systemic causes: poorly designed processes, failed training, pressured decisions. The focus became on what needed to be strengthened and not on who would be punished.

7. This turning point marked the beginning of a new culture that is not perfect, but more conscious, more humane and truly committed to the safety that is lived, not just that which is reported.

RESULTS (AFTER ~2 YEARS):

Increased transparency: The number of reports of near misses has grown significantly and this has been celebrated as a sign of maturity. More reports mean more visibility, more dialog and, above all, more trust.

Living procedures: Procedures are now consulted, discussed and used. Quality and clarity have improved, and feedback from workers has shown that they now make sense in practice.

Leadership present: Being present at the Gemba became part of the routine. Leadership is now seen, heard and felt. Engagement increased, and with it, the connection with purpose.

Latent risks exposed: Problems previously ignored began to appear and, more importantly, to be resolved. Hidden weaknesses came to light before they became work accidents.

Real impact: LTIR fluctuated at first, as expected when the focus shifts from hiding to learning. But high-potential events dropped significantly, revealing a more attentive and prepared system.

Cultural change: Safety is no longer just a goal or an obligation. It is now seen as a shared value, built collectively. A daily choice, lived out in every conversation, every decision, every care. A culture that now breathes more truly.

KEY LESSONS FOR EXECUTIVES:

Beware the illusion: Reactive Indicators, such as a low LTIR, can be reassuring, but also misleading. Good figures don't necessarily mean a safe environment. Compliance on paper does not protect lives if, in practice, the risks remain hidden.

Connect to the real work: Real safety happens in the field, where the work is done. Go to the Gemba. Listen carefully. Observe with respect. That is where the answers are that no report can be provided.

Encourage transparency: An environment where people feel safe to talk is a more prepared environment. Reports of mistakes and near misses are signs of trust and precious opportunities for prevention.

Lead by example: Your presence, your questions, your posture and your decisions, or omissions, communicate. The silent example of leadership is often stronger than any written rule.

Invest in skills, not just paper: Procedures that really help, training that builds real skills, investigations that teach rather than blame all build resilience. This is what sustains safety in difficult times.

Reflection: Could your organization be suffering from the illusion of conformity? What can you, as a leader, do today to bring management closer to the reality of work? What concrete steps can you take to really strengthen the capabilities that keep your team safe?

CASE STUDY 2: THE "TOWER OF BABEL" - COMMUNICATION SILOS AND DISTANT LEADERSHIP

THE SCENARIO:

At Construtora Vertical, a medium-sized company specializing in residential buildings, everything seemed to work properly. There was a structured safety department, regular training sessions were held, and procedures were properly documented. However, behind this apparent organization there was a silent bottleneck, represented by fragmented communication between the various levels and sectors of the organization.

The hidden reality (discovered by a new safety engineer):

Technical "Tower of Babel": The procedures were extensive, technical and not very accessible to those working on site. Language was not enough. In practice, what guided the work was accumulated experience, not what was on paper.

Feedback lost along the way: The voices from the field brought up in DDSs, informal conversations, and attempts to warn rarely reached management. Managers, already overloaded, had no clarity on how or to whom to escalate the information.

"Office" leadership: Management rarely appeared on the construction site, and when they did, the focus was on the progress of the work. Safety was rarely discussed with those who experienced the risks. The feeling was clear: safety was "something for the safety department" and the foremen.

Poor communication with third parties: Subcontractors received initial guidance and manuals, and that was it. There was a lack of continuity, listening and real verification of understanding. They passed on, but there was no dialog.

Fear of "making a case": Workers were reluctant to point out problems. The fear was that they would be seen as hindering the schedule or as "complaining too much". Silence became protection.

Diagnosis: Construtora Vertical experienced a typical scenario of fragmented communication, where each sector spoke its own language and the leadership was perceived as distant from the real safety challenges. A veritable "Tower of Babel" that blocked the flow of vital information and thus undermined prevention, engagement and trust.

THE PAINFUL AWAKENING:

The wake-up call came after a serious near miss while lifting a part that swung dangerously close to the team. The investigation revealed a dangerous combination of factors, including winds above acceptable limits, communication failures between operator and signalman, lack of clear guidance on weather conditions in the procedures, a third-party signalman without a clear warning sign, and the absence of a warning sign. The investigation revealed a dangerous combination of factors, including winds above acceptable limits, communication failures between operator and signalman, lack of clear guidance on weather conditions in procedures, an outsourced signalman without proper training, and radio problems that had been reported but never resolved or brought to the attention of leadership. What's more, right at the critical moment, the direct leadership was absent.

THE KEY TURN:

1. **Clear and visual communication:** The critical procedures were revised with the active participation of masters and workers, with the main aim of making everything more accessible and applicable to everyday life. To this end, visual checklists, simplified instructions and short videos with straightforward language were created. All the content was carefully adapted to the practical reality of the construction site.

2. **Feedback channels that work:** A simple system was created so that managers could register and escalate concerns without noise. A multidisciplinary committee, made up of workers and third parties, began to analyze the reports and propose concrete actions. The flow started to work.

3. **Leadership present and attentive (Safety Walk):** A formal safety walk program was implemented. Managers and directors began to go to the construction site with a new purpose: to observe, listen and talk to those who live the risk. Questions such as "What worries you here?" or "How can we improve this together?" became part of the routine.

4. **Bridge with third parties:** Integration was strengthened. Specific briefings, more direct communication channels and regular meetings focusing on third-party management ensured continuous alignment and not just at the start of the contract.

5. **Psychological safety in practice:** The leadership communicated clearly and repeatedly that reporting deviations, near misses and concerns was not only welcome, but essential. Success stories were shared, showing that those who spoke up were heard and that no retaliation would be tolerated.

This turnaround didn't erase the risks, but it changed the way we see them. The "Tower of Babel" has started to become a bridge. And when people feel that they can speak and that they will be heard, culture really begins to transform.

RESULTS (AFTER ~18 MONTHS):

More useful procedures: Procedures have come to be recognized as real support tools. They have become clearer, more applicable and more present in the site's routine.

Feedback flowing: There was a significant increase in the volume and quality of feedback coming from the front line. Suggestions were not only listened to, but implemented, which strengthened confidence in the process.

Connected leadership: The team's perception of the presence and genuine interest in the leadership has grown. Safety walks stopped being an event and became part of the routine.

Collaboration with third parties: Communication with subcontractors has evolved. There has been more alignment, more listening and more partnership in day-to-day operations.

Better managed risks: With more visibility and action, near misses and minor incidents have decreased, reflecting an environment that is more aware of and attentive to the details that make a difference.

A more open climate: Dialogue about safety has become stronger. People began to collaborate more, trust more and participate more naturally, a sign of a maturing culture.

KEY LESSONS FOR EXECUTIVES:

Communication is vital: Communication failures create risks, even with structured systems. Invest in clarity, accessible language and channels that work both ways. Simplicity connects, complex distances.

Leadership takes presence: It's not enough to lead from afar. Your physical, attentive, interested presence is irreplaceable. Go out into the field, listen with intent, talk less, understand more.

Create bridges, not walls: Safety needs to flow between areas, levels and partner companies, break down silos, bringing planners and executors closer together is a strategic and deeply human gesture.

Feedback is a gift: Even when it brings discomfort, feedback reveals ways forward. See it as an opportunity to listen and evolve. When there is psychological safety, there is room for real learning.

Lead by example: What you do, or fail to do, sets the tone for the culture. Words inspire, but attitudes build. Be consistent, your example is the strongest reference your team has.

Reflection: Is your organization home to "Towers of Babel"? Are there leaders who still live far away from the operational floor? What can you do, starting today, to make safety communication more fluid and the presence of leadership closer, more real, more transformative?

CASE STUDY 3: THE "BROKEN BRIDGE" - MANAGING THIRD PARTY SAFETY

THE SCENARIO:

At Química Industrial Segura (QIS), dependence on third-party companies for critical activities such as maintenance, logistics and shutdowns was high. For its own employees, the management system was robust. With third parties, the requirement was for contractual compliance. On paper, everything seemed to be under control.

THE HIDDEN REALITY (REACTIVE AND FRAGMENTED MANAGEMENT):

Superficial onboarding: Onboarding processes were long, standardized, focused on generic rules and lacked practical context about the real risks of the activities to be carried out. Verification of understanding was minimal: a checklist, not listening.

Delegated/diluted supervision: QIS trusted the contractor to take care of supervision. When QIS appeared, it was to approve permits, without closely monitoring how the work was done.

Reactive communication: The conversation only took place after incidents or during audits. Few spaces to prevent, listen or share lessons learned in advance.

"Us vs. Them" culture: Outsourced teams felt excluded, with less access, less information and less space to be heard. As a direct consequence, distancing, resentment and low collaboration emerged.

Pressure for Deadlines/Costs: At both the contractor and the contracted companies, tight targets trumped safety precautions. Urgency spoke louder than prevention.

Diagnosis: There was a broken bridge. Despite contractual compliance, the real risks associated with third-party work were not being managed effectively. Integration was tokenistic and collaboration fragile.

THE PAINFUL AWAKENING:

A fire that occurred during hot work carried out by an outsourced team during a maintenance shutdown clearly revealed several critical flaws. The risk analysis carried out by the contractor proved inadequate, and the work permit, issued by QIS, had been released without a sufficiently rigorous check. The team was completely unaware of the presence of flammable materials nearby. In addition, there was no efficient communication between the different teams involved.

An event that could have been avoided altogether exposed what many already realized in practice: it is not enough to demand compliance with the rules. It is necessary to build genuine belonging, promote open dialog and guarantee a constant presence. The safety of outsourced teams is also a direct responsibility of the contractor, and no contract can replace the importance of this real, day-to-day commitment.

THE KEY TURN (FROM CONTRACT TO PROACTIVE PARTNERSHIP):

After the incident, QIS realized that demanding contractual compliance wasn't enough. It needed to build a different relationship, one that was closer, more human and more collaborative. The change began with a new management mentality for partnership.

1. **Careful selection (beyond price):** The choice of contractors began to be considered more than cost. An in-depth assessment of the Safety Management System, performance history and, above all, the culture of the partner company was incorporated. The focus has shifted from just "how much it costs" to "how it takes care of it"

2. **Relevant and continuous integration:** Integration has been reformulated. A common basic module, plus specific modules by risk and area of activity. For fixed third parties, QIS opened internal training and created a sponsorship system for someone from the QIS team as a support point for questions, alignment and reception.

3. **Clear roles and responsibilities:** Joint workshops with representatives from QIS and the contractors helped define, in a practical and agreed way, who does what in different scenarios. This eliminated gray areas and strengthened the sense of co-responsibility.

4. **Proactive communication:** Regular meetings were set up between QIS and contractors, with a safe space for listening. In addition, forums were created in which outsourced workers could express concerns without fear of retaliation. Speech gained space and value.

5. **Shared and behavioral supervision:** Audits began to be cross-checked. QIS observed the work of third parties, and the third parties also observed the QIS teams. The focus was on practice, not paper. And the intention was to improve, not punish.

6. **Real inclusion:** Third-party workers were invited to participate in DDSs with QIS teams. Good practices began to be recognized on both sides. The "us vs. them" culture began to dissolve.

7. **Shared indicators:** Proactive indicators of third-party safety are now monitored and discussed openly in joint meetings. What used to be invisible now has a name, a number, dialog and action.

RESULTS (AFTER ~2-3 YEARS):

Strengthened collaboration: Communication between QIS and contractors has become more fluid, frequent and respectful. Dialogue has replaced unilateral demands and solutions are now built together.

Third-party engagement: The outsourced teams began to feel part of the team. Their sense of belonging has increased, as has their active participation in safety-related discussions and decisions.

Proactive risk management: Risks were identified and dealt with earlier. Listening and being present in the field allowed for more effective preventive action.

Fewer incidents: The rate of incidents involving third parties has fallen consistently. The environment has become safer, both in terms of behavior and conditions.

More efficient shutdowns: Maintenance stoppages now have fewer interruptions due to faults or work accidents. Integration and planning have improved overall performance.

Reinforced reputation: QIS is now recognized internally and in the market as a company that values safety for everyone, not just its own employees. An organization that turns contracts into partnerships and care into a culture.

KEY LESSONS FOR EXECUTIVES:

Third-party safety cannot be outsourced: Taking care of the safety of contractors is the direct responsibility of the contractor. It requires presence, involvement and partnership, not just contractual clauses.

Build bridges, not walls: Invest in open communication, with clear expectations and integration processes that make sense to those in the field. Safety is built together and strengthened with trust.

Including to engage: When third parties are treated as part of the team, engagement increases. Belonging generates collaboration. Exclusion generates distance, silence and risk.

Supervision is collaboration: more than monitoring, it is about observing with respect, dialoguing with interest and learning together. Supervision that transforms is one that listens and welcomes, not one that just demands.

Selection matters: Choosing partners based on price alone can be costly. Evaluate culture, history and attitude towards safety. Those who share values build together.

Reflection: How has your organization managed third-party safety? Is it a purely contractual and reactive relationship or is there a genuine effort at proactive partnership? What concrete steps can be taken today to strengthen this bridge and ensure that everyone, without exception, works in a truly secure environment?

CASE STUDY 4: THE "INFALLIBLE" LEADER AND THE NECESSARY CLIMB OF HUMILITY

THE SCENARIO:

Tecno Soluções was a booming company, pulsing with innovation and driven by speed. Its CEO was admired by many as a charismatic, brilliant and respected visionary, and his decisions were rarely questioned. However, behind this inspiring image of genius lay a subtle but extremely dangerous gap. Safety in the labs was never seen as a priority and always seemed small in the face of constant pressure for immediate results.

THE HIDDEN REALITY (LACK OF ORGANIZATIONAL HUMILITY):

Culture of the "infallible genius": The culture that was consolidated within the company strongly revolved around the figure of the "infallible genius". The CEO's view was considered absolute, even on subjects in which he didn't have complete mastery. When he expressed an opinion, no matter whether it was on technology, processes or safety, his words automatically became the rule. Any warnings coming from the teams were usually disregarded and often seen as excessive bureaucracy, resistance or even a lack of strategic vision.

Resistance to change: There was visible resistance to change. As the CEO had designed the company's first processes and layouts, any proposed revision was interpreted as personal criticism, and he reacted. This defensiveness spread like a silent current throughout the leadership.

Safety as a "barrier": Safety? It used to be seen as a brake, an obstacle, a cost. Never as a value, never as part of the journey.

Learning blocked: Small accidents happened - cuts, burns, near misses. But instead of learning, there was silence. Instead of investigation, individuals blame.

It was easier to label it as employee error than to look at the system.

Fear of speaking out (lack of psychological safety): And so, a fear of speaking out gradually set in. Technicians and engineers knew about the risks, saw the flaws, but hesitated. How could they say something that contradicted the CEO? How could they suggest changes if even the managers preferred to keep quiet to avoid friction?

Diagnosis: Behind all that structure there was a simple but deeply rooted problem: the pride presents at the top of the organization, accompanied by an almost total absence of organizational humility.

Decisions weren't shared, people's voices were rarely heard, and learning didn't find enough space to develop. The culture of infallibility stifled collaboration, and where there is no room for vulnerability, there is no genuine safety.

Basically, much more was needed than reorganizing processes or training. Something bolder and more significant was needed: we had to get off the pedestal of perfection and start climbing, step by step, the mountain of humility.

Only in this way would it be possible to build an environment where making mistakes becomes a learning opportunity, where questioning is understood not as disrespect but as a legitimate expression of care, and where safety is no longer seen as a cost but as an essential value.

This climb doesn't start with numbers or targets. It starts with courageous choices, the choice to listen, the choice to review decisions and, above all, the conscious choice to change.

THE PAINFUL AWAKENING:

What was supposed to be just another prototype test ended in smoke, flames and a deep silence that communicated everything. A fire took hold of the laboratory, causing considerable damage. However, greater than the material damage were the truths revealed by the investigation.

There was no single villain or culprit. What came to light was a complex web of failures, including weak procedures, blocked firefighting equipment and critical steps ignored due to time pressure. Many risky practices were already known but had never been clearly discussed. Previous proposals for improvements had been dismissed as "unnecessary costs".

It wasn't possible to pinpoint a single culprit. It wasn't just human error, it was an entire system urgently calling for profound changes.

THE TURN OF THE KEY (THE DIFFICULT CLIMB OF HUMILITY):

The fire was more than an incident. It was an invitation for the CEO to step down from his pedestal and look head-on at the consequences of the culture he helped build.

Faced with this scenario, and with the firm support of an external consultant, the leader who previously knew everything, decided everything and centralized everything had to learn to listen. And that learning hurt. But it was precisely this pain that led to a turning point:

1. **Vulnerability at the top:** For the first time, the CEO publicly acknowledged that the leadership had failed. He apologized to the team for not listening earlier. It was only the first step, but one of the hardest. Admitting to being part of the problem paved the way for becoming part of the solution.

2. **Diverse voices:** The Safety Committee emerged. A space where all voices, from all levels, had a place. From the bottom to the top. The CEO himself sat next to the technicians and operators to listen, not to dictate. Safety began to take on new forms and new voices.

3. **Cultivating psychological safety:** Training came for leaders and managers. The company understood that safety is also emotional, that fear silences warnings and that courage needs space to grow. Anonymous channels were created and, more than that, there was a real commitment that no one would be punished for telling the truth.

4. **Structured learning:** Every incident, every near miss became raw material for evolution. The company has adopted a clear methodology for investigating causes, identifying weaknesses in the system and acting. No longer to punish, but to learn.

5. **"Unlearning" in order to improve:** Perhaps the most symbolic gesture was that of the CEO himself when he gave the committee the autonomy to review and change even the processes that he had once created. Letting go of what no longer served. Learning from the new. Accepting that there are other ways of doing things and that they can be better.

6. **Safety in the targets:** And finally, safety is no longer just a nice talking point, it is now at the heart of the strategy. Proactive Indicators made their way into managers' targets. As a result, safety is no longer "someone's" responsibility, but everyone's.

The pain brought a gift. A gift wrapped in smoke but loaded with the truth that no one leads alone. Culture is not built with command and control, but with listening and presence.

True leadership begins when we have the courage to recognize that we need to change. And this courage, when cultivated, turns pain into learning and making mistakes into a starting point for something much greater.

RESULTS (LONG AND CHALLENGING PROCESS):

It wasn't quick and it wasn't easy, but time showed what the fire had revealed: it was possible to do things differently and better.

Open communication: As listening was strengthened, reports began to come in. Technicians and engineers began to share perceived risks and suggestions for improvement more frequently. Fear gave way to collaboration. The culture of silence began to be broken down, word by word.

Safer environment: Based on interviews and analysis, concrete changes were made. From process adjustments to physical improvements in the laboratories. Safety began to be felt not just in the numbers, but in the details of everyday life.

Fewer incidents: The frequency of minor incidents has fallen consistently. And since then, no serious events have been recorded, not because of luck, but because of awareness.

Leadership in evolution: Managers have come to value feedback. The once centralizing CEO began to rely more on collective intelligence. He delegated; he listened and grew. And with it, the whole culture changed.

Safety as an enabler: From being an "obstacle" to innovation, safety has come to be understood as a bridge, a foundation that allows us to dare responsibly. The future is now built with courage, yes, but also with prudence.

KEY LESSONS FOR EXECUTIVES:

Humility is strength: Pride blinds us to risks. Humility, this silent force, invites us to see, listen, rethink. And change. Admitting mistakes doesn't weaken us. It humanizes us.

Vulnerability generates trust: When a leader allows himself to make mistakes in front of the team, he opens space for others to express themselves truthfully. And this is where the trust that sustains safe cultures is born.

Listen to the front line: Those who execute see. And they see what reports often don't show. To value these voices is to stop managing in the dark.

Error is information: The focus needs to shift from "who made the mistake" to "why was it possible to make a mistake?". That is how you learn. That is how you evolve.

Safety starts at the top: Culture doesn't change with speeches, but with attitudes. It is the example that leads. It is listening that transforms, and it is the willingness to learn that sets the course.

Reflection: Is there room for more humility in the leadership of your organization?

Think: when someone makes a mistake, what happens? When someone speaks, who listens? The way we react to mistakes, suggestions and discomfort says everything about the culture we are building.

Leading with safety is, above all, leading with humanity. Because safety doesn't start with standards. It starts with people.

And people, we take care.

CASE STUDY 5: THE DOMINO EFFECT OF FRAGILE CAPABILITIES - WHEN THE LAYERS CRUMBLE

THE SCENARIO:

Transportadora Rápida was a giant in the logistics sector, with an extensive fleet, nationwide operations, advanced tracking technology, basic safety training and even a corporate department dedicated exclusively to the subject. In their institutional materials, there was a clear message that safety was a priority. However, it only took a slightly closer look to realize that there was something unstable behind that apparent firmness.

THE HIDDEN REALITY (INTERCONNECTED WEAKNESSES):

Safety was in the discourse, but not in practice. A house of cards supported by fragile, disconnected layers.

Governance (Top): Leadership focused on aggressive delivery and cost targets. Safety was mentioned, but resources for significant investments such as advanced training, embedded technologies or expansion of specialized teams were systematically postponed. In the dashboards, proactive safety had little or no weight in the goals of regional managers.

Supervision (Middle): The pressure was on. You had to deliver. You had to hit your targets. You had to do more with less. Long working hours, driving at night or in adverse conditions, forgotten breaks were all tolerated. Safety messages always arrived after something had gone wrong. Management was more reactive than preventive.

Processes (Operational): Vehicle inspections, although mandatory, have become bureaucratic rituals. Papers filled out in a hurry. A "to-do list" that provided little protection. Route planning ignored specific risks and focused only on speed and efficiency.

People (Operational): There was training, yes, but only at the beginning.

Afterwards, there was little reinforcement. The drivers dealt with fatigue, stress and difficult decisions every day. And they knew that talking about it meant delay. It meant questioning. It meant punishment. So, they kept quiet, and drove on.

Diagnosis: The safety said at the top was not sustained at the base. Each layer from strategy to execution had cracks. And, as in a domino effect, one weakness fed another. There was no robustness, there was appearance.

It was a culture where safety was said to be a priority, but where speed was applauded. Where silence about tiredness was interpreted as commitment. Where fear, disguised as productivity, kept everyone moving until something inevitably fell apart.

THE PAINFUL AWAKENING:

It was in the quiet early hours of the morning that it all happened. A truck belonging to Transportadora Rápida was traveling when the driver lost control and the vehicle overturned. The injuries were serious, the cargo was lost, as was the truck.

However, nothing has been lost more deeply than the illusion that everything was under control.

The investigation revealed an uncomfortable truth, clearly showing that the work accident was not an isolated failure. It was, in fact, the cumulative result of several silent weaknesses that had been growing over time. A real domino effect.

Extreme fatigue (people/operational): Long working hours, deadline pressure. An exhausted body trying to meet targets that didn't consider human limits.

Contributing mechanical failure (process/operational): A problem with the brakes that went undetected. The inspection was carried out in haste, as part of a mechanical routine, with no room for real care.

Pressure from supervision: The supervisor knew about the long hours but didn't intervene. The pressure for targets outweighed common sense.

Misaligned goals (governance): Delivery and cost were top priority. Safety? It was in discourse but had no real weight in decisions.

Insufficient training (processes/governance): Fatigue management was superficial. There was a lack of effective mechanisms for a driver to report fatigue or be replaced without fear.

THE KEY TURN (SYSTEMIC RECONSTRUCTION OF CAPACITIES):

The accident was a milestone. Harsh, yes. But it was necessary for TR to begin a profound reconstruction, reviewing all the layers that support safety or that, until then, allowed it to be fragile:

1. **Responsible governance:** The board of directors understands that safety cannot be sustained by good intentions alone. It redefined managers' targets, including proactive safety indicators with real weight. Safety has become part of the company's strategy and budget. It stopped being a declared value and became a visible decision.

2. **Empowered middle management:** Managers and supervisors underwent intensive training, focusing on safety leadership, effective communication, fatigue and the balance between production and care. They were trained and authorized to stop unsafe operations without fear of retaliation.

3. **Robust and intelligent processes:** Pre-trip inspections have been simplified and digitized, becoming more objective and efficient. Route planning now includes real risk analysis. And a robust journey and fatigue monitoring system was implemented, with clear and enforceable rules.

4. **Engaged and trained drivers:** The front line was listened to. Open workshops were held to discuss risks, fatigue and pressure. Fatigue training was no longer one-off, but in-depth and recurring. Confidential reporting channels were also created, offering safety for those who need to say, "I'm not well enough to drive today".

6. **Technology as support (not punishment):** TR began to invest in fatigue monitoring with artificial intelligence and advanced telemetry, not as a tool for punishment, but as support. Technology is now used to identify risks and provide coaching, becoming an ally of safety rather than fear.

RESULTS (COORDINATED AND CONTINUOUS EFFORT):

Reconstruction wasn't quick or easy. But over time, the fruit began to appear. The change took shape, and the results showed that caring does pay off.

Fewer incidents: There has been a significant reduction in events linked to fatigue or mechanical failures. What used to be a risky routine became an avoidable exception. Breaks were respected, inspections were carefully carried out, and warnings were heeded.

Trust and communication: The relationship between drivers, supervisors and managers has changed. Trust has emerged, listening has emerged, talking about safety is no longer a nuisance. It became natural, necessary, welcome.

Adherence to processes: With simplification and clarity, there has been greater compliance with inspections and safe planning. Processes are no longer a burden. They have become protection and support tools daily.

A culture of shared responsibility: Safety is no longer a "function of one sector" but is now seen as a prerequisite for sustainable efficiency. Everyone understood their role. And everyone began to look after each other and the whole.

Sustainable performance: Even with initial adjustments to deadlines, the results have paid off. Greater reliability and fewer incidents have brought stability. And the organization has now been able to continue growing with solidity and awareness.

KEY LESSONS FOR EXECUTIVES:

Safety is systemic: It cannot be guaranteed by isolated actions. Weaknesses at any level can lead to disaster (domino effect). Safety needs to be integrated, structured and aligned.

Governance sets the tone: Its goals, allocation of resources and demands shape behavior at all levels. Safety doesn't happen by chance. It is a direct result of what you prioritize or neglect.

Empower middle leadership: They are the crucial bridge between strategy and execution. They need clarity, support and autonomy to say "no" when it is necessary to protect.

Processes should support people: Simplify, make relevant and facilitate safe working, rather than creating barriers. When processes care, people trust.

Invest in people: Continuous training, listening channels and a culture of genuine care are essential. Especially on the front line, where decisions are made under pressure and on the move.

Reflection: Are the different layers of management and operation in your organization truly aligned in favor of safety, or are there weaknesses and misalignments that can create a "domino effect"?

11

SAFETY HORIZONS

NAVIGATING FUTURE TRENDS, CHALLENGES AND OPPORTUNITIES

As we move through the intense and constant transformations of the 21st century, the scenario in which organizations operate is no longer the same and may never be again. We live in times of accelerated change, where technologies evolve rapidly, work models are constantly reconfigured, relationships between people and companies take on new meanings, and society's expectations become more demanding, more human and, above all, more urgent.

It is no longer enough just to keep up with these changes. It is essential to look deeply, sensitively and purposefully at what is before us - a new way of existing within organizations. As the way we work is being redesigned, it is equally necessary to redesign the way we look after people's safety, health and well-being.

The future of Safety Culture can no longer be seen as a distant prediction or an abstract academic reflection. It is an immediate strategic necessity, an urgent invitation for leaders to develop anticipatory skills, to know how to interpret the signs of change, to act with clarity in the face of uncertainty and to thrive even during chaos.

In this context, an essential question arises. What are the trends, the most pressing challenges and the real opportunities that will shape safety and organizational culture in the coming decades?

Answering this question is not just an intellectual exercise. It is, above all, a concrete gesture of care for the future, the first step in preparing our people, our structures and our values for this new era that is already silent, complex and, at the same time, full of possibilities.

1. THE STRATEGIC CONNECTION: SAFETY AS AN ESSENTIAL PILLAR OF ESG

The ESG agenda, which brings together the Environmental, Social and Governance pillars, has ceased to be a differential and has become a central axis of corporate strategy. What was once seen as a trend is now urgency. And in this new scenario, Occupational Safety and Health (OHS) takes on a prominent role not just as part of the "S", but as a vital link between all the pillars.

Social impact (S): A strong Safety Culture is the most visible way of demonstrating genuine commitment to people. It goes beyond compliance to express care, empathy and responsibility. It has a direct impact on reputation, helps attract and retain talent, strengthens the social license to operate and reduces costs with work accidents, sick leave and turnover. it is, in practice, the tangible reflection of care.

Environmental connections (E): In sectors such as chemicals, energy and mining, process safety failures not only affect people, they also have devastating environmental impacts. Integrated risk management (PSM) acts as a protective shield not only for lives and assets, but also for the environment. Cleaner operations, in many cases, are also safer.

Reflection of governance (G): The way safety is conducted, monitored and sustained within the organization says a lot about its quality of governance. When the Board supervises closely, when there is clarity in executive responsibilities, adequate resource allocation and transparency in reporting, safety becomes a mirror of what the company is about. The demand for ESG accountability today strongly includes OHS metrics.

Trend: Safety is no longer just a technical function or a cost center. It is evolving into an essential strategic component of ESG, linked to long-term value creation. It is not just good practice; it is part of business sustainability.

Implication for leaders: It is time to develop a strategic vision of safety within the ESG, to communicate its value clearly to all stakeholders and to integrate safety into the organization's core decisions. It is time to articulate the business case for safety not just in terms of prevention, but of shared value.

2. THE EVOLUTION OF TOTAL WORKER HEALTH: CARING FOR THE WHOLE PERSON

For too long, we have treated physical safety and occupational health as separate worlds. But real life isn't divided like that. The body, mind, environment and bonds are all connected. it is in this context that the concept of Total Occupational Health is gaining strength, an approach that invites us to take care of the whole person.

Impact of mental health on physical safety: We can no longer ignore the real impact of psychosocial factors on our daily lives. Excessive workload, low autonomy, lack of support, internal conflicts and an imbalance between personal and professional life are all elements that generate constant tension, anxiety and burnout. These conditions do not simply remain outside the company. They enter with each employee, directly affecting cognitive functions that are fundamental to safety, such as attention, decision-making and reaction time, significantly increasing the risks. Fragile mental health also weakens people's physical safety.

Psychological safety as a foundation: If people don't feel safe talking, they keep quiet. And where there is silence, there is risk. Psychological safety is the foundation of physical safety. It enables voluntary reporting, clear communication and collective learning. When there is trust, there is listening. When there is listening, there is care, and it is this care that sustains safe cultures.

From prevention to active promotion: The vision of Total Occupational Health transcends the simple prevention of injuries or illnesses. It is an invitation to actively promote well-being, through effective mental health support programs, consistent stress management initiatives, greater flexibility in the workplace, spaces with real meaning and leaders who don't just guide. It is an invitation to actively promote well-being through effective mental health support programs, consistent stress management initiatives, greater flexibility in the workplace, spaces with real meaning and leaders who don't just guide, but express genuine care in their daily actions.

Trend: Safety Culture is changing. It is evolving into a broader Culture of Care, which does not separate body and mind, but integrates them with sensitivity. Where safety and health go hand in hand, supporting more humane, sustainable and engaged environments.

Implications for leaders: The invitation is clear, adopt a holistic view of well-being, promote psychological safety not as a project but as a living practice, integrate OHS, HR and Occupational Health policies coherently and, above all, led by example with empathy, presence and genuine support.

3. DELVING INTO HUMAN AND ORGANIZATIONAL FACTORS: UNDERSTANDING THE SYSTEM

The realization that human errors are symptoms and not root causes is steadily gaining momentum. Every wrong decision, every apparent failure often points to something deeper, such as systemic weaknesses, misalignments in the organizational environment or invisible pressures present in everyday life.

The discipline of Human and Organizational Factors invites us to look beyond individual behavior. It seeks to deeply understand the interaction between people, the technologies used, the processes implemented and the existing organizational structure. Based on this broader vision, it offers ways to build safer, fairer and essentially more humane environments.

Human-centered design (and the actual task): The proposal here is simple and, at the same time, profoundly transformative. It involves designing tasks, equipment and procedures with careful consideration for people's real capabilities and limitations. The aim is to facilitate safe work exactly as it happens in practice, rather than expecting idealized human beings to continually compensate for imperfect systems.

Proactive fatigue/workload management: Fatigue is not a sign of weakness, but a real and concrete risk. That is why more and more responsible organizations are adopting proactive systems to manage this risk, establishing more humane work schedules, effectively monitoring the physical and mental load of workers and encouraging an environment where reporting fatigue is free of judgment. This care needs to be fully incorporated into the daily routine.

Understanding decision-making: In environments of constant pressure, decisions are made based on multiple intersecting factors. Deep understanding how these decisions are made is essential for improving training, creating concrete support for the moment of decision and perfecting critical processes. When we understand the reasons behind each choice, we can act directly on the factors that influence them.

Consolidating a just culture: There will always be mistakes, that is an inevitable reality. The real question is how we react to them. Just Culture proposes transparent and mature processes to clearly differentiate between three fundamental types of situations:

- Unintentional error, which requires systemic learning.
- Conscious risk-taking behavior, which requires coaching and the removal of barriers.
- And willful violation, which may require appropriate disciplinary measures.

In this way, the focus shifts from individual guilt to a deeper understanding and continuous strengthening of the system.

Trend: Safety is no longer just a set of rules to be followed. It is evolving into a scientific, systemic and human-centered approach. And that changes everything.

Implication for leaders: Leadership needs to promote understanding throughout the organization. Invest in safe work design. Implement Just Culture with consistency. And, above all, see mistakes as a learning opportunity, not a reason for punishment.

4. TECHNOLOGY: OPPORTUNITIES AND CHALLENGES IN THE DIGITAL AGE

Technology has arrived in full force, bringing possibilities that would have seemed distant or even unimaginable a short time ago. Along with it, however, have come new issues, new risks and unprecedented complexities. Faced with this scenario, safety needs to learn to embrace these advances with wisdom, discernment, ethics and, above all, with a deeply human-centered outlook.

The potential of data analysis: The ability to process large volumes of data coming from sensors, reports, operations open a promising window to reveal hidden risk patterns and enable more anticipatory management. But for this potential to be transformed into real care, attention must be paid to the quality of the data, the clarity of the analysis (avoiding relying on "black boxes") and the ethical use of information, always respecting people's privacy.

Sensors and wearables: The growing presence of wearable devices and sensors integrated into the environment, known as IoT, makes it possible to monitor exposure, location, fatigue levels, falls and issue alerts in real time. This is a powerful opportunity for active prevention. However, this technological innovation also requires maturity and discernment. Care must be taken with privacy issues, the sheer volume of data generated and the real risk of alert overload, which can generate noise and confusion rather than clear and effective action.

Automation and robotics: By transferring dangerous, repetitive or extreme tasks to robots, technology can save lives. But on the other hand, there is a need to rethink human-machine interaction, ensure cyber safety, carry out safe maintenance and, above all, take care of the people impacted by automation. The workforce needs to be listened to, retrained and respected. It is not about replacing humans but protecting and enhancing their capabilities.

Virtual/augmented reality (VR/AR): Immersive technologies create safe learning scenarios for high-risk environments. They allow training to be realistic without putting anyone in danger. But their effectiveness depends on conscious investment and careful design with a focus on learning, not just innovation for innovation's sake.

Trend: We live in times of increasing adoption of safety-oriented technologies, which requires a balanced and conscious attitude. We need to welcome these advances with hope, but also with constructive criticism. Innovate, yes, but clearly recognize its limits. Technology needs to be an essential ally of human care, not a substitute for our conscience.

Implication for leaders: Leadership has a responsibility to critically evaluate each technology adopted. The focus needs to be on how it supports people, strengthens the system, facilitates safe working and not just on replacing, controlling or monitoring. Implementation must be ethical, participatory and human-centered. It is essential to invest in digital skills within the safety team, ensuring that advances leave no one behind either technically or emotionally.

Automation, robotics and autonomous systems: Automating tasks that are dangerous, repetitive, physically demanding or carried out in hostile environments can represent a significant advance for safety. Technology has real potential to considerably reduce human exposure to risk, bringing protection where there was vulnerability before.

Today, we are witnessing the growing use of industrial robots, collaborative robots, drones and autonomous vehicles in various sectors, such as manufacturing, logistics, construction, mining and agriculture. Along with these technological advances come significant improvements in working conditions.

However, it is important to remember that every transformation requires special attention. While these systems offer protection, they also introduce new challenges that need to be managed carefully.

It is essential to ensure the safety of the interaction between humans and robots, particularly in contexts where both share the same space. We need to ensure the reliability and cyber safety of these complex systems, carry out maintenance and repair responsibly, and carefully consider the social and organizational impacts that these changes inevitably bring.

In this context, the retraining of the workforce, human acceptance of changes and active listening to legitimate concerns about employment are essential components of this process. After all, no innovation will be truly safe if, along the way, we silence the voices of those who live the reality of work daily.

THE KEY, THEREFORE, IS BALANCE.

For technology to expand care and not just control, it will be necessary to adopt a critical, ethical and human-centered approach from the outset.

Rather than automating for the sake of automating, you must ask:

- How can this technology support human capabilities?
- How can it strengthen the system's defenses, facilitate safe working and enhance the experience of the operator?

The active participation of end users, workers and supervisors in the design and implementation process is what turns technical solutions into practical, useful and well-accepted solutions. Because only those who live the work know what really works.

IN THE END, TECHNOLOGY WITHOUT LISTENING IS RISK. TECHNOLOGY WITH LISTENING IS CARE.

5. ADAPTING TO NEW WORKING MODELS AND DIVERSITY

We are living in a time of profound transformation in the world of work. Globalization, technological advances and new social expectations are redesigning routines, relationships and structures. What used to be an exception, such as remote or on-demand work, is now part of the new normal. And this calls for an honest and agile review of the way we take care of workplace safety and health.

Remote or hybrid work: When the workspace extends to the home, the challenges also move. Ensuring ergonomics in home environments, preventing psychosocial risks such as isolation and difficulty disconnecting, and keeping the organizational culture alive even at a distance are commitments that cannot be neglected. The employer's responsibility does not disappear with the change of address. It changes and requires new policies, adequate support and constant listening.

Non-traditional workforce: Gig economy, outsourced and temporary workers are often in the most vulnerable zones. With less access to training, protective equipment and clarity about who looks after their safety, these professionals face risks that call for new responses. Contracting companies need to rethink contractual approaches, forms of supervision and communication channels, taking responsibility for their commitment to protect everyone who contributes to their operations.

Diversity and inclusion (D&I) in safety: Real safety cannot follow the "one size fits all" model. The workforce is diverse in age, gender, background, skills and this requires intentional adaptations. Policies, training, PPE and ways of communicating need to address the specific needs of each group, ensuring that everyone feels safe, seen and included.

Trend: The buzzword is adaptability. Safety policies and cultures will need to be flexible, fast and inclusive, capable of protecting all workers, in any employment model or workplace.

Implication for leaders: Develop OHS policies suited to flexible working. Take responsibility for the safety of the entire workforce, including those who work in non-traditional ways. And actively promote Diversity and Inclusion (D&I) in safety practices, ensuring that care reaches everyone truthfully, respectfully and consistently.

NAVIGATING THE FUTURE: ADAPTIVE LEADERSHIP, RESILIENCE AND CONTINUOUS LEARNING

If there is one thing we can say with safety about the future, it is that it will be increasingly complex, dynamic and unpredictable. Safety, in this context, is not a fixed destination but a continuous journey, made up of revisions, adjustments and new beginnings.

Navigating this constantly shifting landscape will require more than methods. It will take presence, listening, courage and direction.

Adaptive leadership: The times call for leaders who see beyond the now. They need to be able to anticipate trends, foster innovation and learn quickly from both successes and mistakes. Leading here is also about adjusting routes lightly and quickly, without losing sight of what really matters.

Continuous learning: The organizational culture needs to be open to curiosity, questioning and intellectual humility. It is necessary to sustain an environment where best practices are tirelessly sought, but also where there is the courage to challenge old assumptions, whenever the need arises.

Systemic resilience: Rather than resisting, it will be necessary to build and maintain organizational capacities that allow us to anticipate, monitor, respond and learn in the face of the unexpected. Resilience is not just about technical preparation, but about how the organization chooses to react when the unexpected happens.

Trend: The safety journey has no finish line. It is alive, perpetual and constantly reinventing itself. It requires constant movement, continuous learning and a willingness to change whenever necessary to protect life.

Implication for leaders: More than simply keeping up with change, it is essential to actively lead the process of adaptation and learning. Resilience must be cultivated as a strategic capacity, far beyond a one-off reaction. At the same time, it is essential to always focus on what really matters: valuing life, exercising coherent leadership by example, promoting genuine engagement with people and ensuring that the learning generated by the system returns to the system itself, generating positive transformations around it.

PRACTICAL ADVICE:

Integrate safety and ESG: Position safety as a strategic part of the sustainability agenda. It is not a separate item; it is one of the pillars underpinning commitment to the value of life.

Invest in well-being: Mental health and psychological safety are the real foundations of performance and protection. Taking care of people is the first step to protecting results.

Think systemically: Errors are not just individual. They often reveal weaknesses in the system. Promote a fair culture and build environments where learning is possible even in difficult times.

Adopt technology with caution: Technology should support people and strengthen the system. It shouldn't replace our monitor. Use it ethically, clearly and always listening to those on the front line.

Adapt to the new times: Review your OHS policies with an eye on remote working, flexibility and the diversity of the workforce. Care must keep pace with reality.

Lead adaptation: Be a living example of continuous learning and resilience. In times of change, your ability to listen, adjust and inspire will be your greatest asset.

QUESTIONS FOR REFLECTION:

1. Is your organization truly prepared to integrate safety more strategically into the ESG agenda, or does it still occupy a peripheral place in central decisions?

2. How can your leadership actively promote a "Culture of Care" that values not only physical safety, but also people's mental health and emotional well-being?

3. What concrete steps can your organization take to deepen the application of the HOF and Just Culture principles? Is error still viewed with judgment or with a desire to learn?

4. Is there a structured plan to evaluate and adopt new safety technologies in an ethical, conscious and human-centered way? Is technology at the service of care?

5. Do your company's OHS policies reflect the new times? Do they embrace flexibility, remote working and the real diversity of the workforce?

SAFETY IS NOT A DEPARTMENT, IT IS EVERYONE'S RESPONSIBILITY, STARTING WITH YOU. TAKE THE HELM OF THIS TRANSFORMATIVE JOURNEY.

12
THE SICKNESS OF CULTURE

THE ILLUSION OF CONFORMITY AND THE IGNORANCE OF OPTIMISM

In the long and demanding marathon of cultural transformation in safety, it is not always external obstacles or a shortage of resources that trip us up more often than not. Often, what really paralyzes us are those silent "illnesses" that settle internally, invisible to the eye but deeply damaging to the organizational culture.

These subtle, almost imperceptible states slowly undermine our progress and keep us trapped in constant cycles of stagnation, even when we believe we are doing well and when the indicators look stable. Two of these internal conditions, perhaps the most dangerous, are the illusion of conformity and the ignorance of optimism.

The illusion of compliance occurs when all the forms have been filled in, the training has been carried out and the figures appear to be within the targets set. However, in practice, people remain disconnected from the true purpose of safety. The formal process is followed, but real care is not experienced.

Ignorance of optimism, on the other hand, happens when we believe that everything is fine simply because no serious incidents have occurred recently. In this case, silence is mistaken for stability, and the absence of work accidents is mistakenly taken as evidence of a safe culture. As a result, we stop investigating deeply, listening carefully and looking beyond the obvious.

Understanding and diagnosing these internal conditions is essential for any organization that truly wants to build a robust, resilient and lively safety culture. Because these conditions reveal an even deeper issue: a failure to perceive reality, a disconnect between what we believe is happening and what occurs daily, on the ground and in the details of people's behavior.

As long as this disconnection is not addressed with lucidity, courage and concrete actions, no external effort will be sufficient to sustain a true and lasting transformation.

THE SILENT DISEASE: ILLUSION OF CONFORMITY (WHEN PAPER DOESN'T REFLECT PRACTICE)

Among the most dangerous pitfalls on the road to safety maturity is one that is disguised as progress: the so-called illusion of compliance. It arises when we believe that having properly documented procedures, training carried out, audits approved, and apparently positive indicators means that we are on the right track. We feel a comfortable sense of accomplishment, even when genuine care is not present in everyday practice.

This illusion arises from the confusion between superficial compliance and real safety, between what looks complete and perfect on paper and what happens daily, in the real workplace. When we look at organizations that have undergone successful cultural transformations, such as the experience documented by McKinsey with the company Total, we see that the real change happened when safety stopped being treated as a priority and began to be experienced as a fundamental corporate value.

COMMON SYMPTOMS OF DELUSION

Celebrating long periods without recorded accidents as if they were incontrovertible proof of safety is one of the most obvious symptoms of this disease, zero LTIR becomes a trophy, but by only looking in the rear-view mirror, we forget the near misses that are ignored, the latent risks and the silent complexity of real workplace safety.

Blind trust in audits is another worrying symptom. Passing an audit becomes synonymous with "we're safe", but little thought is given to whether the audit reflected real day-to-day life or just a stage set for the occasion.

Procedures as an end in themselves, what we might call a "paper fetish", also reveal this illness: we value the formal existence of procedures, their technical formatting and their appropriate location in the system. We often forget to ask ourselves if these procedures are applicable in practice, if they are clear enough, if they are accessible to those who need them and if they are alive and active in everyday operations. If the answer to these questions is no, then what we have is not a real ally, but just a stored file.

Training sessions such as "checklist" are yet another symptom of this condition: the workload is met, the annual plan is filled out, but no one asks whether the content was absorbed, whether there was a change in behavior, whether it made sense to those who were there; you learn the minimum to pass, not enough to protect.

The hidden danger: The illusion of compliance is especially treacherous, as it lulls the leadership into a false sense of tranquility, sustained by the belief that "we're fine, the figures show it". As a result, risks begin to accumulate silently, while vigilance weakens and necessary investments end up being postponed under the justification that "we are already compliant".

In this scenario, workers clearly see the disconnect between what is said and what happens daily. Faced with this incoherence, feelings of cynicism, disengagement and disbelief begin to emerge. The organization becomes complacent not because it doesn't care, but precisely because it believes it is safe. it is precisely in this misperception that the danger lies: when the apparent comfort of formal conformity leads us to neglect the true essence of safety.

Let's remember the emblematic case of a procedure that took 16 minutes to access. On paper, everything seemed correct and adequate, but in practice, at the critical moment when it really mattered, it proved to be completely useless. This is the illusion of conformity in action, safety that exists on paper but protects no one when they need it most.

THE EMPTY OPTIMISM TRAP: OPTIMISTIC IGNORANCE

Alongside the illusion of conformity, we find another equally silent trap that is often driven by the best intentions: the so-called optimistic ignorance or empty optimism. This type of optimism arises from the legitimate desire for positive results, but is unaccompanied by a real commitment to deep, consistent actions that are fully connected to the practical reality of work.

This optimism is not linked to the real effort that change requires, to listening sincerely to people and to a clear understanding of how transformations really happen within organizations and their teams. When we analyze success stories in cultural transformation, such as those documented in resilient organizations that make psychological safety a strategic priority, we realize that genuine optimism is always accompanied by concrete and sustained action.

SYMPTOMS OF OPTIMISTIC IGNORANCE

Acute "Toolitis", characterized by the excessive implementation of tools without real engagement, often manifests itself in seemingly modern and promising initiatives, a new reporting system, an innovative spreadsheet for risk management, or even a newly launched behavioral program. It all sounds advanced and full of potential, but if there is no active listening to people, effective training, an environment of trust and the removal of practical barriers, these tools end up being just props.

The illusion is created that their mere existence is enough to bring about change, when they haven't even begun to be experienced daily, it is like planting seeds in arid soil and expecting them to grow just because they've been put in the ground.

Unilateral communication, or "talking to the walls", is another obvious symptom: bulletin boards are full, emails are sent, slides are shown repeatedly, but no one really listens, no one feels involved or part of the process. When communication happens in a one-way, impersonal way and without the effective involvement of the leadership, it completely loses its power and relevance.

Talking about safety doesn't necessarily mean communicating it; without open dialog, real listening and concrete, visible actions, words become empty, incapable of generating any real change; it is like shouting in a valley and hoping that the echo is enough to move mountains.

Passive and disconnected "training" also reveal this illness. Expecting a change in behavior from a generic video or a procedure given to you to read alone is naïve, to say the least; safety needs to be experienced, discussed and practiced. When learning is passive, without interaction or connection to reality, the content does not become care.

Setting unrealistic goals, what we might call "wishful thinking", completes this symptomatic picture: ambitious targets, tight deadlines, limited resources. When we try to implement on paper what the system, people and culture are not yet prepared to support, the inevitable result is frustration. To assume that simply pushing for targets will be enough to bring about genuine transformation is to confuse the will to change with the actual capacity to do so.

THE DEEP DIAGNOSIS

The ignorance of optimism exposes a significant gap in the understanding of how real change happens, it starts with the mistaken assumption that simply telling people what needs to be done or launching a new initiative is enough to automatically transform everything.

This kind of thinking completely ignores the complexity of human behavior, the weight of productive goals, the decisive influence of context, the need for practical skills, the crucial role of active listening and, above all, the vital importance of leadership that is truly present and deeply committed to the process.

THE DEVASTATING CONSEQUENCES

When we act out of empty optimism, not only do we fail to achieve the expected results, but we end up eroding something even more valuable: trust. Time is lost, resources are wasted, and what's worse, people start to lose faith, with each new tool launched without a real purpose, with each goal imposed without proper support, the silent thought arises that, yet another idea has arrived that won't last.

With this feeling comes cynicism, disengagement and withdrawal. This is because people quickly realize when an action is superficial or disconnected from practical reality. What was born as an opportunity for progress ends up becoming yet another point of disbelief on the road to change.

It is as if each poorly planned initiative is a small betrayal of people's trust, an unfulfilled promise that gradually erodes the credibility of the leadership and the willingness of teams to embrace future transformations.

THE ANTIDOTE: REALISM, HUMILITY AND FOCUSED ACTION

Overcoming the illusion of conformity and the ignorance of optimism is no simple task. But it is undoubtedly an indispensable step for any organization wishing to build a Safety Culture that goes beyond paper and truly touches people's lives. This overcoming begins with a genuine commitment to realism, humility and intentional action, and requires the courage to look inwards deeply and the willingness to act consistently.

1. SEEKING THE TRUTH ON THE FACTORY FLOOR

It is essential to go beyond formal reports and meeting rooms, to be present where the work really happens, to observe carefully, to listen with genuine interest, to talk directly to the people who carry out the tasks, to understand the "little ways", the adaptations made on a daily basis and the challenges faced on a daily basis.

The Gemba, a Japanese term meaning "the real place", is where the truth comes out unfiltered, it is where you see what paper doesn't show, it is where the antidote to illusions lies, because nothing replaces contact with the reality people live. When leaders are willing to leave their offices and walk around the workplace, not just to inspect, but to genuinely understand, they discover a wealth of information that no report can capture.

This practice goes far beyond an occasional visit, it is about establishing a consistent routine of presence, where leadership becomes part of the day-to-day operations, not as a watchdog, but as a partner in building solutions. It is in this direct contact that the real barriers emerge, the unmapped risks, the creative adaptations that keep the operation running and, above all, the real opportunities for improvement.

2. MEASURING CAPABILITIES, NOT JUST RESULTS

Developing and monitoring Proactive Indicators represents a fundamental change in perspective, quality of leadership, usability of procedures, reporting culture, effectiveness of training, not just looking at work accidents that have happened, but at the signs that something might happen.

Reactive Indicators show what has already happened; Proactive Indicators allow us to see what can still be avoided. They reveal the health of the system, while low numbers of work accidents alone can hide latent risks, it is like the difference between measuring only a patient's fever versus monitoring all the vital signs that indicate their general health.

Mature safety organizations develop metrics that capture the vitality of their culture, the frequency and quality of safety conversations, the response time to concerns raised, the percentage of people who feel comfortable reporting near misses, and the effectiveness of corrective actions implemented. These indicators paint a much richer and more reliable picture than simply counting days without work accidents.

3. FOCUS ON THE QUALITY OF IMPLEMENTATION

Less haste, more depth, prioritizing the quality of delivery, co-creating with users, communicating clearly, practical training and ongoing support, creating real conditions for change to happen, removing barriers and facilitating implementation.

A good idea that is poorly implemented generates frustration and when this happens again, people stop believing in it. Better one initiative that is well done and thoroughly implemented than many that are poorly understood and quickly forgotten, it is the difference between planting a tree carefully, watering it daily and monitoring its growth, versus scattering seeds in the wind and hoping that a few germinate.

The quality of implementation manifests itself in seemingly small but fundamentally important details: the time devoted to explaining the "why" of a new practice, the patience to listen to doubts and resistance, the willingness to adjust to the course when something isn't working, the care to celebrate small progresses along the way. The patience to listen to doubts and resistance, the willingness to adjust course when something isn't working, the care to celebrate small advances along the way.

4. CULTIVATE ORGANIZATIONAL HUMILITY

Building an environment where it is safe to admit mistakes, ask questions, raise doubts, question "it has always been like this", recognizes that we don't know everything and that learning needs to be continuous at all levels.

Humility is fertile ground for true transformation. When an organization believes it already knows everything, it closes the door to the new, shuts out voices from the front line and loses the chance to evolve. Truly resilient organizations, as studies on psychological safety have shown, create spaces where vulnerability is seen as a strength, not a weakness.

This organizational humility is expressed in many ways: leaders who admit when they don't know something, teams that celebrate those who have had the courage to report a mistake, systems that treat failures as learning opportunities, cultures that value the right questions more than quick answers.

5. ACT ON THE BASIS OF REAL DIAGNOSES

Making decisions based on in-depth, contextualized and realistic diagnoses, eschewing generic solutions, hasty copies and market fads, looking inwards before looking outwards for answers.

Effective solutions are those that respect the context. Copying a model without understanding its roots is like taking medicine without knowing the name of the pain - it may relieve it for a while, but it is unlikely to cure it. Each organization has its own history, culture and unique challenges. What worked for one company may not work for another, even if they both operate in the same sector.

Real diagnoses require time, method and, above all, the courage to see the truth, even when it is uncomfortable. They involve multiple perspectives, combine quantitative data with qualitative insights, and always consider the broader context in which the organization is inserted.

6. HEALING AS A CONTINUOUS PROCESS

By recognizing and actively treating the illusion of conformity and the ignorance of optimism with this set of antidotes, we redirect our energy to where it really makes a difference, we stop chasing indicators and start strengthening real capacities, we exchange the effort of keeping up appearances for the commitment to protect people.

But it is important to understand that this healing isn't a one-off event, a moment of enlightenment that solves everything at once, it is an ongoing process, a constant vigilance, a daily choice to prioritize substance over appearance, depth over superficiality.

As the experience of organizations that have succeeded in truly transforming their Safety Cultures teaches us, the path to healing requires persistence - Total, for example, took years to transform safety from a priority into a genuine corporate value. It wasn't an overnight change, but the result of consistent action, committed leadership and, above all, the courage to look inwards and recognize where the real weaknesses were.

7. THE ROLE OF LEADERSHIP IN HEALING

In this healing process, leadership plays a fundamental role, but not as the holder of all the answers, but as a catalyst for transformation. Leaders who manage to overcome these cultural illnesses are those who develop the ability to ask the right questions, to create safe spaces for honest dialog and to demonstrate, through their actions, that safety really is a non-negotiable value.

These leaders understand that their main responsibility is not to have all the solutions, but to create the conditions for the solutions to emerge from the collective knowledge of the organization; they know that the people who carry out the work on a daily basis have valuable insights into the real risks and opportunities for improvement.

Effective leadership in curing these ailments manifests itself in specific behaviors such as being present at the Gemba regularly, asking open and genuine questions, listening without judging, acting on what has been heard, communicating decisions transparently and, above all, showing vulnerability when appropriate.

8. THE IMPORTANCE OF STRATEGIC PATIENCE

One of the most important lessons in the process of curing these ailments is the need to develop what we can call "strategic patience" - the ability to stay focused on the long term, even when the pressures of everyday life push for quick, superficial solutions.

Organizations that manage to overcome the illusion of conformity and the ignorance of optimism are those that develop the discipline to invest time and energy in changes that may not show immediate results, but that build solid foundations for the future.it is like the difference between taking a painkiller for a headache versus treating the underlying causes that are causing it.

This strategic patience doesn't mean passivity or slowness; on the contrary, it requires constant and targeted action, but always with the awareness that profound cultural transformations take time to consolidate. It is the difference between running a 100-meter race and running a marathon: both require speed, but completely different strategies.

9. REACHING THE FINISH LINE: CULTURE AS AN INFINITE JOURNEY

Step by step, we're approaching the finish line of this cultural marathon, not in a hurry, but with consistency, clarity and a kind of strength that you can't see on paper, but that you feel daily, a culture that sustains itself not because it seems safe, but because it really is.

But herein lies a paradoxical truth about this journey: the finish line of cultural transformation into safety is not a destination, but the beginning of a new stage. When an organization reaches the interdependent level of cultural maturity, when "we take care of each other" becomes a lively reality and not just a slogan on the wall, it doesn't stop evolving, on the contrary, it becomes capable of continuous and adaptive evolution.

10. THE RESILIENT CULTURE OF THE FUTURE

Looking at the horizon, we can glimpse what the safety culture of the truly resilient organizations of the future will be like. These organizations will not just be those that have managed to eliminate work accidents, but those that have developed the ability to constantly learn, adapt and evolve in the face of new challenges and changing contexts.

The Safety Culture of the future will be characterized by some distinctive qualities, firstly, it will be deeply human, recognizing that behind every procedure, every decision and every action there is a person with their own motivations, concerns and aspirations. This culture will not treat people as resources to be managed, but as partners in building a safer and more sustainable environment.

Secondly, it will be adaptive by nature; instead of relying rigidly on fixed rules and procedures, it will develop the ability to respond intelligently and creatively to new and unforeseen situations. This does not mean abandoning standards and guidelines but rather creating a culture that knows when and how to adapt these standards to the specific realities of each situation.

Thirdly, it will be deeply connected to the organization's higher purpose, safety will not be seen as a cost or an obligation, but as a fundamental element of operational excellence and business sustainability. Third, organizations with mature Safety Cultures will find that genuinely investing in people's well-being not only reduces risks, but also improves quality, productivity and innovation.

11. THE ROLE OF TECHNOLOGY AS AN ALLY

As we move into the future, technology will play an increasingly important role in the evolution of safety cultures, but it is crucial to understand that technology alone will never be the solution, it will be a powerful tool in the hands of organizations that have already developed solid cultural foundations. Will never be the solution; it will be a powerful tool in the hands of organizations that have already developed solid cultural foundations.

Artificial intelligence, advanced sensors, augmented reality and other emerging technologies can significantly amplify the ability to anticipate risks, monitor conditions in real time and respond quickly to dangerous situations.

But these technologies will only be truly effective in organizations where people trust each other, where there is psychological safety in reporting problems and where the leadership is genuinely committed to everyone's well-being.

Technology of the future will not replace the human element in safety, but will enhance it, allowing people to focus on what they do best, thinking critically, making complex decisions, looking after each other and adapting creatively to new situations.

12. COLLECTIVE RESPONSIBILITY

One of the most striking characteristics of mature Safety Cultures is the realization that responsibility for safety is truly collective. This goes far beyond the traditional concept of "everyone is responsible for safety", it is a responsibility that is lived, felt and actively exercised by every person, at every level of the organization.

In these cultures, safety becomes a natural expression of mutual care, people don't follow safety procedures just because they must, but because they genuinely care about the well-being of their colleagues, their families and their communities, they understand that every safe action is an act of love and respect for others.

This collective responsibility manifests itself in countless ways, in the courage to stop an activity when something doesn't seem right, in the willingness to share knowledge and experiences, in the initiative to propose improvements, in the patience to teach and learn, in the humility to admit mistakes and in the determination to always look for better ways of doing things.

13. THE LEGACY WE HAVE BUILT

When we look back on the entire journey of this cultural marathon, we can ask ourselves: what kind of legacy are we building? What kind of organization are we leaving for the next generations? What example are we setting of how it is possible to work safely, productively and humanely?

The legacy of a mature Safety Culture goes far beyond the numbers of work accidents or awards received, it manifests itself in the lives that have been preserved, the families that have remained intact, the communities that have benefited from responsible and empowering operations. But perhaps the most important legacy is the demonstration that it is possible to create organizations where people feel truly valued, cared for and empowered.

This legacy is perpetuated when people who have experienced organizations with mature Safety Cultures take these values and practices elsewhere, they become ambassadors for a different way of working and living, spreading seeds of transformation wherever they go.

14. THE JOURNEY CONTINUES

We've reached the end of this book, but not the end of the journey. In fact, if you've made it this far, it is likely that your journey is just beginning or entering a new phase. The ideas, concepts and tools presented throughout these pages are just the starting point for a transformation that will be unique to your organization, your team and your specific context.

The Safety Culture marathon has no definitive finish line, it is a continuous journey of growth, learning and evolution, each day offering new opportunities to strengthen the culture, take better care of people and build a safer and more sustainable future.

But remember, you're not alone on this journey. Around the world, thousands of leaders, safety professionals, workers and organizations are taking similar paths. Every small step you take, every improvement you implement, every life you protect, contributes to a global movement towards a safer and more humane world of work.

Cultural transformation in safety is ultimately an act of hope, it is the belief that we can do better, that we can create organizations where no one gets hurt, where everyone feels valued and where work is a source of fulfillment and not suffering. It is the belief that we can do better, that we can create organizations where no one gets hurt, where everyone feels valued and where work is a source of fulfillment and not of suffering.

This hope becomes a reality through concrete actions, committed leadership, engaged people and, above all, through the courage to keep going, even when the road seems long and challenging, the journey continues, and it continues with you.

15. PRACTICAL ADVICE FOR THE JOURNEY

- Be wary of "calm"

- When the numbers look too perfect, it is time to take a closer look, zero work accidents don't always mean zero risk, ask yourself "What are we doing, proactively, to sustain this result?" and "How do we know people are feeling safe to report?"

- True safety is not silent, it is characterized by active conversations, constant questioning and a collective vigilance that never completely relaxes, truly safe organizations are those where there is always something being discussed, improved or questioned in relation to safety.

- Go to the Gemba regularly

- Get down from the office, go to where the real work happens, observe, listen, talk without judgment and ask with genuine openness "What prevents you from doing your job safely?" and "Do our procedures help or hinder?"

- Make it regular practice, not an occasional visit, establish a routine of presence that allows you to really understand the nuances of daily work, and remember, you are there to learn, not to inspect.

- Question quality, not just quantity

- More important than the number of initiatives is how much they impact practice. When evaluating any action, ask: "Have people been engaged?", "What has changed in the routine since then?" and "Is this being lived or just implemented?"

- Develop metrics that capture the depth of the transformation, not just its breadth. it is better to have one well-implemented and sustained change than ten initiatives that are only on paper.

- Celebrate humility and learning

- Mistakes and near misses are windows to growth, when a problem is reported and dealt with responsibly, celebrate it, recognize those who had the courage to speak up, value learning as a collective achievement, show that error is not to fail, it is to learn with truth.

- Create rituals and practices that reinforce this mentality, share learning stories, publicly recognize those who contribute to continuous improvement through their observations and suggestions.
- Demand diagnosis before solution
- Before saying yes to a new tool or action, firmly ask "What real problem are we trying to solve?" and "Do we have evidence that this solution fits our context?"
- Develop the discipline of always thoroughly understanding the problem before looking for solutions, invest time in rigorous diagnoses, resist the temptation to copy solutions that have worked elsewhere without first understanding whether they apply to your specific context.
- Stay focused on your purpose.
- Never lose sight of the fundamental "why" of all this effort: protecting lives, preserving families, building a safer and more sustainable future. When the pressures of everyday life threaten to divert your focus, always return to that central purpose.
- Communicate this purpose regularly, tell stories that connect daily actions with the bigger impact, help people see how their individual work contributes to something much bigger than themselves.
- The journey of cultural transformation in safety is, above all, a journey of courage, courage to look at the truth in the face, even when it is uncomfortable, courage to act, even when the path is not completely clear, courage to persist, even when the results are slow to appear.
- But it is also a journey of hope, the hope that we can create better, more humane and safer organizations, the hope that every life is protected, every work accident avoided, every family preserved makes all the difference.
- And it is, finally, a journey of love, love for the people we work with, love for the families who depend on us, love for the future we are building together.
- The marathon continues and it continues with you, one step at a time, one day at a time, one choice at a time.

- Far Beyond Zero, towards a culture where everyone gets home safe, every day.
- Because without diagnosis, there is no direction. Without action, there is no transformation. Without persistence, there is no culture.

Made in the USA
Monee, IL
13 November 2026

d1ab2f6a-473f-4742-98db-2638bdba387bR02